# Creative
# Mobile Media

## A Complete Course

# Creative Mobile Media
## A Complete Course

Sylvie Prasad

*University of East London, UK*

**World Scientific**

NEW JERSEY · LONDON · SINGAPORE · BEIJING · SHANGHAI · HONG KONG · TAIPEI · CHENNAI · TOKYO

*Published by*

World Scientific Publishing Europe Ltd.

57 Shelton Street, Covent Garden, London WC2H 9HE

*Head office:* 5 Toh Tuck Link, Singapore 596224

*USA office:* 27 Warren Street, Suite 401-402, Hackensack, NJ 07601

**Library of Congress Cataloging-in-Publication Data**
Names: Prasad, Sylvie E., author.
Title: Creative mobile media : a complete course / Sylvie E Prasad,
    University of East London, United Kingdom.
Description: New Jersey : World Scientific, [2017] | Includes bibliographical references.
Identifiers: LCCN 2016055440| ISBN 9781786342805 (hc : alk. paper) |
    ISBN 9781786342812 (pbk : alk. paper)
Subjects: LCSH: Mobile computing. | Multimedia systems.
Classification: LCC QA76.59 .P697 2017 | DDC 006.7--dc23
LC record available at https://lccn.loc.gov/2016055440

**British Library Cataloguing-in-Publication Data**
A catalogue record for this book is available from the British Library.

Desk Editors: Chandrima Maitra/Mary Simpson/Shi Ying Koe

Typeset by Stallion Press
Email: enquiries@stallionpress.com

For Udayan, Aria and Nicholas

# About the Authors

**Sylvie Prasad** is a visual artist, senior lecturer in Media and Communications at the University of East London, UK and a senior fellow of the Higher Education Academy. She has published on lifestyle television and celebrity culture and exhibited work exploring the metropolitan experience and everyday photographic practices. More recently her work has utilised mobile phone technology to create three short films. *Who are Ya?* examined Arsenal football fandom, belonging, masculinity and mobile phone culture. *May Days* was an autobiographical 'mobile-doc' and focused on how the phone-camera was used to create a 'short-term memory' for her mother who suffered from Alzheimer's. The work was featured in 'Pocket Cinema' BBC Radio 4, UK and 'Pick of the Week', BBC Radio 4, UK. *Office no. 47* returned to themes of belonging and how the working environment gets transformed into personal spaces through the use of images and artefacts. She was one of the first academics in the UK to create a course for undergraduates that explored the specifics of using mobile phone technology for creative practice and currently leads the module Mobile Media, Exhibition and Promotion at UEL.

**Helen Powell** is a Principal Lecturer and Subject Head in the School of Social Sciences at the University of East London. Helen has written widely on the subjects of advertising and consumer behaviour and is currently working on the 4th edition of *The Advertising Handbook*

(Routledge). Her interest in time and temporality commenced with her PhD which was later published as *Stop the Clocks! Time and Narrative in Cinema* (I.B. Tauris). Helen adopts an interdisciplinary approach in her teaching, writing and research drawing particularly on literature, film and promotional culture.

**Max Schleser** is a filmmaker, who explores innovation in digital film-making and creative media production through wireless, mobile and ubiquitous technologies. His portfolio includes various mobile, smart-phone and pocket camera films which are screened at film festivals internationally. He is the Creative Director of collaborative digital storytelling platform *24 Frames 24 Hours* (www.24frames24hours.org.nz). He co-founded MINA, the Mobile Innovation Network Australasia (www.mina.pro) and curates the annual International Mobile Innovation Screening as well as produces the International Mobile Innovation DVD and eBook. He is also a Senior Lecturer in Film and Television at Swinburne University of Technology, Melbourne, Australia.

**Gabriel Moreno Esparza** is a lecturer in Journalism in the Department of Media and Communication Design at Northumbria University. Gabriel is involved in the delivery of under- and post-graduate modules such as Writing for Publication, Classic Journalists, Digital Journalism, History of Journalism and Research Methods.

**Anandana Kapur** is an award-winning filmmaker and co-founder of Cinemad, India. Previously, an executive producer for non-fiction programming on TV, she has worked on information and video campaigns for GOI and UNICEF.

# Acknowledgements

Thanks go to World Scientific for commissioning the book and for Alice Oven who first suggested there was a book to be had. Many thanks to Mary Simpson at WS for keeping everything on track and seeing it through to publication. I would also like to thank my brother William Whitfield who encouraged me to write and who shared the experience of having a mother with Alzheimer's. I am indebted to the Alzheimer's Society and the wonderful support they give through their informative website likewise the resources found through the Selfie Research Network. Thanks go to Arsenal FC and the helpful people at reception and all those answering my endless telephone queries. Thanks especially to the Arsenal 'lads'. I would not have been able to complete the book without the help of numerous colleagues at the University of East London. Special thanks go to Professor Andy Minion and Ajay Choksi at the Rix Centre. Both spare time each year from their busy schedule to run a workshop on my module. Thanks to my dear friend and colleague Helen Powell, for keeping me sane and for your contribution to the book. To Max Schleser, once a student on my production modules at UEL — what a pleasure to see your academic career go from strength to strength. Thanks too, go to your chapter collaborators Gabriel and Anandana. The book wouldn't have been possible without the support of all my students. This book is really for all of you. Special thanks go to Nikki Mills and her mum Sue, to Serife Akinci and Ramandeep Punhani for the use of your poster, to Annie Alta, Georgina Styles, Shaiann

Mangan and Anis Davanizadeh and to all past students in my Mobile Media class. Finally, a special thanks to my family; for Aria and all your patience while I get my best shots; Nicholas, for giving me such generous access to your mobile phone clips, text messages and your help with content on football, fans and all things Arsenal. Last but never least Udayan, for your really useful suggestions, proofreading skills that are second to none and for your continuing love and support.

# Contents

# List of Table and Figures

# Introduction

'... the mobile phone is portable, to the extent of seeming to be
an extension of its owner, a personal object constantly there, at
hand ... individuals seem to carry their network of connections
which could be activated telephonically at any moment.' (Liccope in
Elliott and Urry, 2010: 48)

I am standing at the Tower of London on a sunny July morning
watching tourists take photographs of each other against the historical backdrop. They can be divided easily into two groups. One group
are using expensive digital DSLR cameras with large detachable
lenses, the rest, the majority, are using their mobile phones, some
attached to selfie sticks. I make the journey back to my home in
north London by tube and in the packed carriage become aware of
how few the passengers that are not using a mobile phone in some
shape or form are. To journey through a city by underground, bus,
or on foot you cannot go very far without noticing the headphones
or the rapid movement of thumbs and fingers as people listen to
music, message each other or scan their phones heads bent downwards. Not confined to the cityscape similar scenarios can be witnessed across towns, countryside and on holiday beaches. The
argument that people do not talk to each other face-to-face anymore
might be over-played but glancing round any restaurant will locate at
least one couple or family engaged not with each other but with
some activity on their mobile phone. Since the first telephone call

was demonstrated using a mobile phone in 1973 its technological development has been phenomenal. The latest smartphones, with camera, sound recording and Internet connectivity are now multimedia devices. They have been credited by some social commentators as changing the face of contemporary cultural practice at the centre of a technological revolution embedded in everyday life. More cautious claims position the smartphone, at the very least, as a 'remediator', that is, existing media practice is repositioned through the new technology (Goggin and Hjorth, 2014). Debates such as these have highlighted challenging and sometimes contradictory positions that have informed part of academic study and scholarly interest around both technology in general and in more recent times the specifics of mobile media. More populist commentaries covering similar themes are to be found online and in newspapers.

John Naughton writing in *The Guardian* newspaper on the threats of technology and biosciences to liberal democracy makes a pertinent comment when he suggests: 'Most writers in the implications of new technology focus too much on the technology and too little on society's role in shaping it' (2016). Naughton's article taps into some of the concerns articulated in the first half of the 21st century that ask questions about where we see ourselves as humans in the future. Some of the more dystopian predictions have envisaged a world where machines, Artificial Intelligence and algorithms will have ultimate control. Others have questioned the effects of digitisation on the human brain to ask if technology is altering our relationship to reality and the implications behind this (Self, 2016). There is no denying the fact that we are entering a new era in which the pace and scale of technological change has far exceeded anything experienced in previous eras. Some have even called for the naming of a new epoch, the Anthropocene, in recognition of the influence of human activity on the planet (Zylinska, 2014). But it is perhaps the uncertainty of where such developments may take us, amplified through links to a new information and data-driven culture, that has created much of the current anxiety. In considering the mobile or smartphone's contribution to this new technological age this book is also turning to the big questions that seek to understand the relationship between individuals, society, technology and cultural practice. Naughton's statement implicitly acknowledges the tensions between the techno-determinists who foreground technology as the driver in shaping the world and those who see a human hand shaping the way technology is developed and utilised. As we begin to examine the significance of mobile media and the smartphone as a tool for communication and creative practice, some of these ideas might start to frame our own thinking when undertaking a practical course of study, discovering along the way the scope of technology and its role in re-shaping the cultural landscape.

Photography emerged around the beginning of the 19th century. Previous discoveries in chemistry and optics were combined and 1839 is usually given as the standard date for the birth of photography's more practical form. Liz Wells (1998/2015) in *Photography: A Critical Introduction* urges caution in any singular claims but suggests it is more useful to examine it as a 'set of practices' along with the particular 'contexts' of its production. As it evolved, each new technological achievement offered the possibility of a different way of both looking at and engaging with the world. The box Brownie of 1900 was the first camera to bring the practices of photography within the reach of the ordinary person. Later, Kodak's Standard and Super 8mm film and the home video cameras did the same for the moving image. The Leica and Erminox cameras of the 1920s and 1930s, combining a streamlined shape with powerful lenses, were linked to the development and expansion of documentary practices to the *auteur* photographer and as the forerunner to the growth of *Street Photography* as an identifiable genre. The sense of immediacy and spontaneity engendered through the technology meant the media produced was understood as: 'intrinsically "modern" and "democratic" ' (Bate, 2009: 47).

So what makes the mobile phone camera so distinct? We might argue it is nothing more than an extension of the small compact digital video cameras or that it can be used for the types of photography the Leica was rightly famous for. The mobile phone differs in one respect. It is the technology we are most likely to carry at all times making it uniquely accessible. As Liccope (Elliot and Urry, 2010) suggests, its portability is 'seeming to be an extension of its owner.' Always at hand are the resources for taking photographs, making films, recording sound, writing, texting and perhaps lastly, the possibility of making and receiving phone calls twenty-four hours a day. The latest smart-phones come with a connectivity that allows for multiple and almost instantaneous dissemination across social media sites on a global scale. It can be argued therefore, the mobile phone's emergence is part of a continuum of technologies that extend our capacity for communication. Ideas of communication lie at the heart of this book; how particular modes of communication can be utilised for the benefit of community and society. In designing a complete creative course the aim is to extend the practical and theoretical knowledge of the reader within a framework of accessible technologies. Whilst not every aspect of mobile media is covered, the focus is on its most popular usage — namely photography and filmmaking — and its role as a social tool.

The book is designed primarily for undergraduate students and follows a particular approach used in the delivery of a final year module across two terms as part of my teaching. It is complete in the sense it forms a self-contained module of study. It is flexible enough to allow additions and modification and tutors may see it more as a guide rather than a prescribed set of tasks and readings. I have attempted, particularly in the first section, to make the reading easy and accessible and the language, where possible, plain and simple. I have been guided very much by my students who are keen to grapple with complex ideas, don't want to be talked down to, but *do* sometimes struggle in their early studies with some of the more dense and impenetrable texts of academia.

For students, the book is divided into two distinct parts. Part One is a practical, *how to* guide, for producing and exhibiting mobile media which encourages a self-reflective critical methodology. At the end of each chapter they will be able to engage with a choice of practical, skill-based exercises and where appropriate link work to a newly expanded global community of mobile media producers. Part Two contextualises practice within a framework of current media, communications and cultural research relevant to mobile phone technology. Specifically, this will extend and underpin knowledge gained from undertaking the practical work in Part One and give them some experience of different approaches and styles of writing.

On my own course we start with the ideas explored in the first chapter of Part One: *Getting Started*, then dip in and out of selected readings as appropriate. My message to students is this: there is no rigid order in which to do things and if your interest lies more in working with the moving image then this is as good a place as any to begin your study. However, be mindful of your approach and try to extend your skills beyond what you already know and like. Be open to new ideas and challenges. By completing the course you should have developed a skill-base and key research methods to create your own chosen media project within a critical framework and have a foundation for moving into postgraduate studies or a more practice based environment. At the very least, it is hoped you will think about how the most common of technologies can become powerful tools in communicating ideas beyond the personal. The following gives you an overview of what you will find in each chapter of the book.

*Getting Started* introduces the reader to the possibilities of creating media via their mobile phone. It will encourage them to be proactive and *write their own manual.* They will focus on specifications relevant to the technology they own and be guided through this exploration completing a table of contents and applications provided in the book. They will be introduced to key terms such as: format, aspect ratio, HD and Applications (apps). They will look at how the limitations of technology can often be the starting point in producing creative work. They will be given tips on how to brainstorm ideas based around personal interests as well as being encouraged to develop work which becomes outward facing; moving from the 'me' to how ideas can be shaped for distribution to a wider audience and ultimately have meaning for others. As such, the scope of this book is to go beyond a *how to do* manual and encourage a critical, reflexive approach to producing mobile media projects.

*Mobile Aesthetics and Criticism* focuses on aspects of creative photography, moving image work, sound and the archive and how basic story telling is a foundation for creating a shared experience. We also examine how the voice memo function on smartphones can be used to capture, for example, particular family histories or moments in time: this introduces the reader to the possibilities of creating social archives such as BBC Radio 4's *Listening Project* (2014). In documenting the changing landscape of mobile media and its specific properties of immediacy and discretion, students will initially be guided through traditional formats. More specifically, however, they will be shown how to utilise camera phone technology for creating unique and meaningful projects and encouraged to engage with new forms of hybrid media and experimentation. In gaining an understanding of current thinking around aesthetics we engage with what it means to be a critic, to analyse your own work and that of other media producers from a framework of informed opinion.

*Promotion and Exhibition* introduces students to the range of available platforms for the distribution of work. The annual iPhone and Mobile Phone film festivals and the Mobile Innovations Network Australasia (MINA) create physical and virtual spaces for screenings and networking on a global scale. Social media sites such as YouTube, Vimeo and Vine create wide possibilities through their distribution channels. At a more local level, exhibiting in a 'pop-up' gallery or producing work for a charity may be more appropriate for some practitioners. The chapter will also deal with what it means to create projects for an audience, keeping work legal by understanding copyright laws and the variety of ways of getting permission to use visual and audio material into the public domain.

Dr Helen Powell begins Part Two with an investigation into Elliot and Urry's (2010) concept of 'networked individualism.' In her chapter *Always On: Mobile Culture and its Temporal Consequences*, Powell analyses to what extent 'we develop personalised nuanced, multilayered temporal registers.' She explores the changing relationship to time itself with clocks and watches, timepieces of modernity, giving way to the centrality of the mobile phone in a postmodern age. She suggests of the mobile phone: 'not only do these devices at the macro level allow for a global temporal reach but also inform at the micro level our sensibilities at any particular moment. How we feel in time is shaped by an "always on" connectivity that aligns with a Deleuzian notion of "duree": of every moment being shaped and informed by what has gone before and what might be.'

*May Days: An Examination of Mobile Filmmaking, Family and Memory*, is the first of two chapters that foreground the theory/practice relationship of creating media within a critical framework of enquiry. How does mobile phone use and filmmaking in terms of the mobile-documentary, contribute to a wider understanding of a relationship with the lived experience? In this particular case, Burgin's notion of 'sites of memory' (2004) are key to creating an understanding of the mother–daughter bond where the mobile phone becomes a therapeutic tool in the creation of memories for someone living with Alzheimer's disease.

*Who are Ya?: Football, Masculinity and Mobile Documentary* is the second chapter to take the practice of creating work on a mobile phone as a starting point to explore questions about the lived experience. Here, identity, masculinity and what it means to be a football fan are investigated. Not least, it also considers if the limitations of the early model of camera phone was behind the creation of a new aesthetic in creative mobile media.

*The Selfie and Multimedia Advocacy* takes as its starting point the current media (and some academic) preoccupation with 'the selfie' as an indication of a more narcissistic, individualistic society shaped increasingly by our relationship to a celebrity-driven culture. It suggests, however, through a study of what it means to document the self via social media, that a more complex narrative is emerging and which, ultimately, links us back to questions about what it is to be human. The chapter continues by discussing the use of mobile phone photography and the *selfie* through the work of Professor Andy Minnion at the Rix Research Centre, UEL, UK. Here, pioneering work is being undertaken in the context of debates about representation, disability and identity. The chapter concludes with an approach to creating student projects that critically engage with the *selfie* as a mode and means of connectivity and media advocacy.

Dr Max Schleser, Dr Gabriel Moreno Esparza and Anandana Kapur co-author the chapter *Mobile Journalism: Reflexive Storytelling in the Co-produced Public Sphere*. In 2015, the British Broadcasting Company offered a training course in the use of mobile phones for filmmaking and photography basics for its journalists recognising the increasing role this kind of technology would play in newsgathering. Schleser, Esparza and Kapur investigate how the mobile phone has contributed to broadcasting and that *citizen journalism* is at the fore-front of many aspects of global news stories. They continue with case studies from Mexico and India, specifically mobile content dissemi-nated via YouTube, which has triggered public debate, and consider if mobile journalism (mojo) — specifically user generated content — and the more 'reflexive forms of story-telling' are part of a new digital environment that has the potential for creating social change.

*A Question of Ethics,* the final chapter of the book seeks to raise awareness of the complexities around creating ethnographic, docu-mentary and photography projects where the human subject is the object of analysis. It examines particular case studies by way of illus-tration and encourages students to focus on their responsibilities as photographers, filmmakers and creators of media projects along with their relationship to the global distribution networks and communi-ties the Internet offers.

My message to students is that I hope by the end of this book you will feel confident and that by engaging with a range of practice and readings offered here, you will have an understanding of the kinds of creative projects that can be achieved with a mobile phone. Thinking critically about the rapidly changing media landscape and how new technology and the Internet can be utilised to expand knowledge, creativity and social engagement is a high ideal but well worth cultivating. I will end this introduction with the words of John Hegarty, the founding Creative Partner for the advertising agency BBH: 'If confidence is one key to success, enjoying your work is another. Even more than confidence, the sense of excitement that accompanies being creative will spur you on' (2014: 3).

# Part One

# Chapter 1

# Getting Started

This chapter introduces you to the possibilities of creating media via your mobile phone. You are encouraged to be proactive in your approach and *write your own manual*. The chapter focuses on specifications relevant to the technology you own and guides you through a detailed examination. You will be introduced to key terms such as: format, aspect ratio, HD and applications (apps). You will look at how the limitations of technology can often be the starting point in producing creative work. You will be given tips on how to brainstorm ideas based around personal interests as well as being encouraged to develop work which becomes outward facing; moving from the 'me' to how ideas can be shaped for distribution to a wider audience and ultimately have meaning for others. As such, the scope of this book is to go beyond a 'how to do' manual and encourage a critical reflexive approach to producing mobile media projects.

We are in a transitional phase with no clear pathway indicating how the digital world may evolve in the future. Multi-digital platforms, hybridisation, layering and shifts in how we understand time have started to influence media practice. Ritchin (2013) compares digital signals, pixels and a code-based media approximating a quantum universe that will require new ways of thinking. The ambitions for this course is to understand how analogue practice has shaped our universe but be open to and prepared for exciting and meaningful transformations in a quantum world. So, with this in mind let us begin our exploration.

Photography has been described as 'painting with light' (Jacobi and Kingsley, 2016) and it is worth remembering that it is in the manipulation of light and how it registers on the surfaces of the subject or object that creates the photograph. Light is transmitted through the lens of your camera phone and converted into a digital signal. Most smartphones automatically calculate the available light levels for you to take what are considered to be optimum quality pictures. The sensor, shutter, aperture and focal length work to mimic the functions found in conventional cameras, particularly the less sophisticated point and shoot devices. If you have had a mobile phone for some time you have probably been using the camera phone and are already familiar with some of its features. It is recommended that you carry out some of the following investigations as a reminder of the functions and limits of your personal handset. This section of the book also details what have become common 'operating conventions' across a range of media practices (Bate, 2009: 1). In this respect, the aim is to start thinking of how the technical manipulation of your tools enables you to see and 'capture' the world and get the kind of images and film that has some meaning. The initial exploration of using your camera phone is about how existing conventions work, or not, with the new technology. In that sense some of the first exercises are very much a development and extension of analogue practice. It is also important to highlight that digital media has already transformed the world we live in and continues to do so.

The simplicity of using mobile phone technology is part of its appeal. It is first and foremost an accessible medium and the aim of this section is to utilise and build on this fact so it becomes a tool amongst many you can draw on for creative practice. The technical descriptions have been kept to a minimum but some details of what you can do with your phone and how it can be used should make for an enjoyable discovery.

## 1.1  Writing Your Own Manual

Technology changes quickly and providing a *how to guide* based on current available mobile phones would most likely be out of date by the time this book goes to print. For this reason, it is important that you start to understand the equipment you will be working with and have some general criteria that enables you to gather knowledge that is relevant to your personal experience. By writing your own manual you will start to explore some of the features on your phone and how you might use its various functions in producing relevant work. It will also make you aware of limitations and how projects may be shaped to meet the challenges encountered. Whilst you may have read the manual that comes with your particular handset, the table of contents (Table 1.1) will help you to focus on the media aspects of your equipment and to familiarise yourself with each of the functions. Some of the checklist contains specifications from smartphones currently on the market. Use it as a starting point to compare with your own handset. At the end of the exercise you can jot down any new specifications or improvements. Camera functions and lens quality are regularly updated but sound quality has tended to lag behind in this respect.

You should have your phone in front of you as you go through the exercise but you can also use the Internet and any instruction booklets to gather further information. Make a note of what each function does or can be used for. As you develop your work you can refer back to this as a checklist.

## 1.2 Apps and Equipment

Some traditional photographers and filmmakers may shun the use of post-production apps and their apparent gimmickry and quick fixes. If you can get an app to alter your colour balance, style, remove unwanted sections and add a nostalgic filter why bother to learn the relevant skills so the argument goes. The answer with creative mobile media is that both are worth exploring. It is always better to take the best possible images you can, to care about composition and start to understand how light and shade affect the photograph. It is better to get the best possible sound recordings and filming and the planning and information in *Getting Started* is there to help create a mindset that puts these principles first.

Table 1.1   The mobile manual: Table of contents.

| Phone make | Notes/function |
| --- | --- |
| Camera pixel size | |
| Aperture | |
| Lens size | |
| Exposure controls | |
| Image stabilisation | |
| Auto HDR for photographs | |
| Face detection | |
| Photo geo tagging | |
| Panorama mode | |
| Flash | |
| Video function | |
| HD video recording | |
| Video frame rate | |
| Video resolution | |
| Video stabilisation | |
| Time lapse | |
| Slo-mo and frame rate (fps) | |
| Zoom function | |
| Audio recording | |
| Memo function | |
| Sound recording with photography (Sound and shot) Audio formats | |

However, I shall also be encouraging a critical look at practice. Since the emergence of photography and film in the 19th century, there has always been a tension between purists and those who embrace new technology. People who placed painting at the top in the hierarchy of the visual arts saw the intrusion of photography as something of a gimmick. When photography became commonplace there was a separation between fine art photography, documentary and domestic or more personal photography.

The late Jo Spence (1934–1992) critically engaged with these contested spaces and common assumptions and put the domestic and personal at the heart of much of her practice. She used the techniques of the snapshot and brought this into the gallery space, converging the notions of art and the everyday. Her body of work broke with popular conventions of what 'art' and documentary photography should be and raised issues about practice to include gender politics, class, family, health and the body (Spence, 1995). To link this to a discussion on mobile phone technology and whether or not to use post-processing via apps or any other software, may seem a tenuous connection but it does highlight the relationship between technology, usage, conventions and attitudes. The emergence of any new technology and how that technology gets utilised is shaped by past experience and knowledge. This can mean opposition to, or reinforcement of, past traditions and the relationship between those that have power to define or contest new practices (see Chapter 7 and the selfie for further reading). So it is worth looking at apps, considering their function and purpose alongside the work you want to produce. Be prepared to ask yourself difficult questions. Why, for example, might it be appropriate to use a sepia filter? Does this change the meaning of the images? Are you trying to recreate a feeling of nostalgia? Does this work in the context you want the images to be viewed? Could your interpretations of nostalgia as a postmodern irony be read as naive by another audience? By considering a rationale for your approach grounded in research you are in a better position to be open to new ideas and make informed choices. You might be a pioneer in pushing the boundaries and forging a new aesthetic but this can only come through knowledge of existing practices and a critical look at the context of production. With this in mind some common apps and their uses are worth exploring.

There are two main kinds of apps (Bamberg *et al.*, 2011). The first is used during the production stage and enhances the qualities of the existing camera making it more sophisticated. It usually gives you an additional range of functions and controls. These are the most useful as they can turn your phone camera into a more powerful tool. The second kind of app is utilised in post-production to enhance or manipulate the image after it has been captured by the camera. Try out some of the most popular free apps. They can be fun to experiment with. There are also useful apps you have to pay for but check if you really need these before making any purchase. It is a good tip to read both professional and user reviews rather than relying on app store descriptions. Much can be achieved using the in-built technology of your phone and you may find enhancements are not needed for the work you want to produce. If you are following this book as part of your university course then you may well have access to industry standard software such as *Adobe Photoshop*® or professional film editing software for post-processing work. New apps regularly come on the market but I've selected some below that have been around for some time and have wide appeal.

It would be difficult to speak about photo manipulation and photo-sharing without mentioning *Instagram*. It was co-founded by Kevin Systrom and Mike Krieger in 2010 and is now owned by Facebook. It is perhaps the most ubiquitous of the photo-sharing apps and well used by students and indeed, also by celebrities. Photographs taken on a smartphone can be manipulated using the app's pre-set filters then shared across online platforms such as Facebook, Twitter and Flickr. The Instagram community number some 300 million users and it is thought 60 million photos are shared daily across the globe (https://www.instagram.com/about/us/, accessed January 2016).

*Hipstamatic*® remains a popular photography app and taps into the nostalgia for analogue techniques and aesthetics (see Fig. 1.1).

Fig. 1.1  Original image and with Hipstamatic filters added.

Photojournalist Damon Winter used it for its aesthetics but this was a rather controversial choice for a series of photographs taken in Iraq and published in the *New York Times* in 2010 (Winter, 2011). Hipstamatic utilises the mobile camera phone and has a wide range of filters to alter the look of the photograph and recreate digitally the style of vintage cameras including a square format. The app has the ability to upload images to social network sites and is part of its appeal but that feature has now been somewhat overshadowed by the development of Instagram. The updated version of the app for the iPhone has a 'Pro' mode that enables adjustments to focus, white balance, exposure and shutter speed. There are numerous other apps on the market that are designed for manipulating images. This kind of post-processing has created a new art genre and following on social media and sites like Flickr and Tumblr have popularised the sharing of the manipulated photograph. There is something to be gained in the fun and freedom of being able to play with images without the seriousness attached to other types of genre photography. More information can be found at http://hipstamatic.com/camera/ (accessed January 2016).

*Snapseed®* for android and iOS marketed by Google is a popular free photo editing app that had a major update in 2015. It has a range of special effects and filters offering a professional level tool that is easy to use. By tapping on the question mark button information is presented in the form of short animations making instructions easier to follow. It features non-destructive editing that allows you to keep original versions of photographs. The latest version includes new tools such as a Lens Blur and a Perspective Transform tool. Further information is available at https://support.google.com/snapseed/?hl=en#topic=6155507 (accessed June 2016).

*VSCO Cam®* for both android and iOS from the Visual Supply Company not only has the usual photo filters or presets but also editing tools that are said to appeal to the more serious photographers. Filters digitally recreate the look and feel of once popular professional film stocks such as Kodak Ektachrome 64, Agfa Ultra 100, Fuji Sensia 100 and Ilford Pan F Plus. There is much nostalgia for creating photographs that reproduce the look and feel of film. Further information is available at http://www.vsco-cam.com/ (accessed June 2016).

*FILMic Pro®* is a popular mobile video app turning the mobile camera into a high definition (HD) video camera. It comes highly rated by professionals and was used on a smartphone to shoot the feature film *Tangerine* (2015, USA). More information is available at http://www.filmicpro.com/apps/filmic-pro/ (accessed May 2016).

*Audioboom* (see Chapter 2) is an iPhone and android app that provides easy access to a hosting service which it publicises as: "Listen, Create and Share". Radio/audio channel or podcasts can be created and material uploaded directly from a mobile phone. Institutions such the BBC and national newspapers have used it to expand online projects alongside individuals wanting to set up niche radio content.

### 1.2.1 *Equipment*

The much maligned selfie stick can be a useful aid in keeping your phone steady. When filmmaker Cleo Barnard made her first mobile film *Dark Glass* (2006, UK) for *Single Shot* she had to attach her mobile phone to a hockey stick with tape to film her scene. How much easier this would have been if the selfie stick had been invented back then. There are two kinds of selfie sticks. The simplest form is nothing more than an expandable rod with camera mount that acts like a monopod in giving you both distance and some stability. The second works via a Bluetooth connection setting with your phone. It enables you to control the shutter release button from the handle of the selfie stick.

Macro lenses are an interesting addition to the tool kit. They fit over the lens of the mobile phone and allow for extreme close up photography. Like all equipment they vary in quality so read the reviews before you purchase anything.

As mentioned previously, sound recording on mobile phones is the least sophisticated of the functions and poor sound recording can ruin many a good student film. There are a range of small microphones that can plug into the headphone jack on the phone and improve the quality of the recordings considerably. Some journalists favour microphones that slot into the charging port of a smartphone and can often be seen in use during live news interviews on television. It is worth remembering to set smartphones to *airplane mode* or *call divert* when recording sound so that any incoming calls don't interrupt the recording process.

The range of apps and equipment you can find to increase the capabilities of the smartphone are endless. You need to decide how useful some of these really are for the kind of project work you want to do. This book is a celebration of the mobility and connectivity of the smartphone so if you find yourself needing too may add-ons this may not be the medium for you.

## 1.3 Composition and the 'Rule of Thirds'

Professional photographs and filmmakers follow particular common conventions. Through time these have been established as 'rules' to follow that by and large improve work or help create particular meaning for an audience. It is by their repetition that they have become established and to an extent, determined what is classified as a 'good photograph'. Rules can be broken and creative mavericks are often at the forefront of setting new codes of construction and new aesthetics. However, what separates them from beginners is that they generally know what the established conventions are and if they choose to ignore these they have a creative rationale for doing so. Sometime a new aesthetics emerges as a way of overcoming the challenges and limitations of a particular technology (see Chapter 6 in Part Two).

The *rule of thirds* is a common compositional device based on proportion and placement of elements within the frame of the image. It is utilised in most visual media including stills photography and moving image (see Fig. 1.2). Aligning the placement of the main subject matter with the lines on the grid is thought to be most aesthetically pleasing to the eye. It can also strengthen the composition.

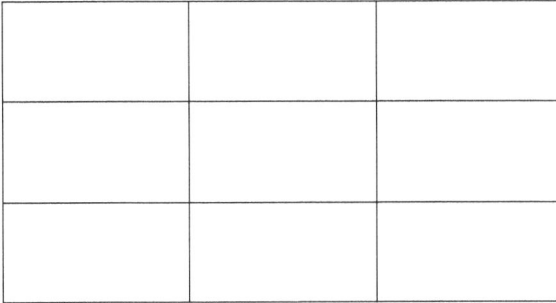

Fig. 1.2 The rule of thirds.

The latest smartphones usually have a grid setting you can select when using the photography or filming modes. This grid setting can be used as a guide to compose your shots and doesn't appear on your final images. Horizons are usually adjusted to be either below or above the centre line but rarely placed in the middle. In filming an interview, for example, most professional camera operators follow these rules. There are exceptions to this you may already be aware of. Consider the mug-shot and its function. This is not meant to be aesthetically pleasing but simply to provide as much detail as possible. Similarly, television newsreaders are usually positioned in the centre of the screen. To reiterate, there are exceptions to most rules but for the most part it will be appropriate to compose your work with consideration to the *rule of thirds*. Through repetition you will be following institutional practices, which have come to signify particular ways of looking at the world and at the same time become familiar with the exceptions. Building up this knowledge and understanding puts you in a better position to critique or challenge, should you feel it appropriate, the current aesthetics.

The orientation of images and how you hold your phone to enable you to compose the best images, also needs careful consideration. Tradition in photography dictates one of two ways when you are using a rectangular screen with *landscape format* and *portrait format* the most common compositional framings. However, these names do not necessarily tie you to only taking landscapes in the horizontal or portraits for the vertical alignment. It is about choosing the best composition for the subject matter and the context the images are to be displayed in. Experiment and see what works best for what you want to achieve. The rise of modernism after the First World War heralded a wealth of experimentation with angles, close-ups and abstraction as part of the exploration and representation of a new modern world. This period saw the birth of movements such as surrealism, constructivism and the avant-garde (Wells, 2015). There is a good chance that your experimentations will have been done by someone else at some point in time but researching photography's rich history will give you an understanding of where your own work 'sits' in relation to what has gone before.

Film and video has always been screened in a rectangular format with a specific ratio between the frame height and width. This is known as the *aspect ratio*. Creative mobile media was much less restrictive and early work produced was often a reflection of the size and the format of the devices used. The early Blackberry model of mobile phone led to some unique square and grainy moving images and created a particular aesthetic. As the quality of technology has improved some of this aesthetic has been lost with a return to a more conventional framing. The aspect ratio of your film is an important consideration when submitting work to festivals. If in doubt or if you haven't decided where you want your work to be shown it is best to film horizontally. There are web channels that may require particular formats with notable exceptions to the horizontal rule. The first work for the *24 Frames 24 Hours* project (Schleser and Turnidge, 2013 and Chapter 2) required a portrait format so three films could be lined up together and played simultaneously across a horizontal screen.

## 1.4 Resolutions and Image Size

Mobile phone cameras are designed for ease of use and require little technical knowledge as many of the setting are automatically controlled. It may be a case of simply setting a square or rectangular format for still photographs or deciding whether you want to use the HD, setting for film. Ease of use has been part of the appeal of mobile phone cameras and that has led to an expansion in photography and filmmaking at the domestic and non-specialist level. Transferring images and films to social media sites is also relatively simple with much of the work accomplished by automated settings. It is only when you start to use more sophisticated editing software or want to print or resize photographs that some technical understanding becomes useful in getting the best results.

A pixel is the smallest single component an image is made up of. Pixel shapes can be square as in a photograph or rectangular as in film. Image quality is measured in pixels and pixel size is initially determined by the quality and size of the lens setting and the camera's capabilities. The resolution can also be thought of in terms of the amount of detail the image holds and current mobile phones have improved the quality of cameras considerably. From your checklist you may have discovered that the camera on your phone is an 8 Mega Pixel (MP) (8 million pixel) or even a 16-MP camera. A 2-MP photograph gives an image 1200 × 1600 pixels which is adequate for publishing to a website (Bamberg *et al.*, 2011). For printing images, particularly if you want to go beyond a very small size, you will need a camera phone with a high MP camera, of the kind that now comes as standard with smartphones. If you are intending to produce work billboard size then a mobile project is not appropriate. It is worth considering, even at these early stages, what the final output of your project will be. For example, will it be screen based, perhaps via the Internet or do you intend to print the work for display or create a photographic book?

### 1.4.1 *Pixels Per Inch (PPI) and Dots Per Inch (DPI)*

There has been some confusion between the terms PPI and DPI not least because they have been used to represent the same things in some accounts of images resolution. The Internet is full of attempts to correct such mistakes with numerous blogs expanding on the various accounts and definitions. Below is a simplified version.

PPI refers to the number of pixels in an inch of the image. A camera set to a high resolution allows for the creation of images with a greater number of PPI. The PPI affects print size and quality of output. If there are too few PPI they are large and create what is termed a pixelated image. That is, the individual pixels can be noticed. Therefore, the higher the number of pixels, the better the quality of the image. A general standard for good quality prints is usually 300ppi and above. This means you will need a minimum of 300ppi source pixels in your original image (Bamberg *et al.*, 2011).

DPI is derived from printing and is relevant if you intend to print your final images. Pixels are made up of colours that in printing get matched physically to the coloured inks in a printer. They are created by a series of tiny dots of coloured ink and are measured as DPI. The higher the DPI the more tonal range you are likely to get in the final printed image. A printer set to 1200 dpi would use 1200 dots of ink per inch and a 1440 printer, 1440 dpi.

### 1.4.2  *Aspect Ratio*

The aspect ratio for an image or screen is the relationship between the width and height. This is commonly written as two numbers. A square photograph, for example, would have an aspect ratio of 1:1. In stills photography, the aspect ratio have been carried over to the digital formats from their pre-digital counterparts.

Common photography sizes are:

1:1 square based on pre-digital medium format cameras,
3:2 based on 35 mm film size of the pre-digital SLR cameras now common for DSLR cameras,
4:3 based on point-and shoot camera and the first computer monitor sizes,
16:9 HD.

Moving image sizes:

1.33:1 or written as 4:3 standard TV,
1.78:1 or written as 16:9 HDTV.

Information on Aspect Ratio is available at http://calculateaspectratio. com/photo-video-aspect-ratio (accessed April 2016).

At this stage it is important to start to put some of the knowledge you have acquired into practice by completing the three simple exercises in photography, moving image and sound. This will allow you to practice composition and framing, control of movement and to assess the quality of your sound recording. Learning through practice is key to any progression in media production and you need to allow time at the end of the exercises to evaluate both the process and results.

---

**Practical 1: Photography**

Using the camera mode turn on the grid in the settings.
If your camera doesn't have this function you will have to visualise the grid in Fig. 1.2 as you compose your shots.
Landscape or portrait format can be used as appropriate.
Using the 'rule of thirds' you will create five photographs that are linked to a specific theme.
One of the images should be a portrait of a person and one image should be a close-up of an object or part of an object. Think about how you might use pattern, texture or shape in the other three images.
Before you press the shutter check your composition and framing.

Reviewing your work:
Identify which photograph is your strongest and consider why.
Did you get all your images in focus?
Was there anything distracting in the frame?
How was the light falling on your subject?
Was the image too dark or too light?

## Practical 2: Moving Image

Use the video function on your phone and set the grid on the settings if you have this function. If not, you will have to visualise the grid in Fig. 1.2 as you compose your shots.

Landscape format should be used.

You are going to film somebody pouring a drink into a glass or making a cup of tea in three shots each lasting a maximum of 30 seconds. You will not use any sound but simply concentrate on the visuals. You should film each shot from a static camera position. This means you will hold the phone as still as possible. The only movement should be in the shot itself.

One of your shots should be a close-up.

Reviewing your work:

Are the shots following a logical sequence?

How might they be edited to make a coherent sequence? For example, where might you put the close-up in the sequence?

How steady did you manage to keep the phone?

Think about the light in each shot. Is anything too dark or too light?

Did you consider the position of the phone and how you framed the shots? Try shooting the shots from a low or high angle. Experiment with angles to see how this changes the meaning of the sequence.

## Practical 3: Sound

Using the sound or memo function of your phone you will make three sound recordings each to a maximum length of 30 seconds. Remember to set the phone to *airport mode* or *call divert* so the recording won't be interrupted.

The first recording should use the phone in a hand-held mode and can be of you reading something or describing an object.

(*Continued*)

*(Continued)*

The second recording should capture a conversation between two people. Place the phone on a table between the two subjects and try and use a quiet space or room to do this.

The third recording should be in an open space to record ambient sound. This could be a stream or water or traffic noise on a busy street.

Reviewing your work:
Consider which sound recording was the most successful and why?
Could you identify any unnecessary background sounds?
Were all the conversations audible?
Was there any distortion in the sound?

You may find that all the recordings had very poor sound quality and this was the least satisfying exercise. As stated before, this has been the least sophisticated of the mobile functions but there are ways of overcoming some of the limitations and these will be discussed in further chapters.

## 1.5  Editing

As mentioned previously, most smartphones are designed for the non-specialist users. You are able to perform a number of limited editing function including cropping and changing the colour balance in stills photography or trimming film clips in video. For more sophisticated editing you will require an app or access to editing software on a computer. Consulting the manual for your particular device will provide the easiest way to upload your images and film to a computer. Some technology has *Bluetooth* connectivity while others require USB cables. At undergraduate level professional edit packages such as *Final Cut Pro-X*®, *Premier*®, *Avid*® for moving image, *Adobe Photoshop*® for still images, *Logic*® for sound tend to be standard part of skills training. Exactly what professional software packages are available to you is determined by what the university or college has access to. They sit alongside many of the useful free software — such as *Audacity*®, *iMovie®* — and a variety of free apps that are available as standard on computers or for downloading online. It is not within the scope of this book to cover detailed aspects of the use of any specific software. Have a look at the online user recommendations as a guide and any free packages that come with your computer. You will find many of the free software packages are perfectly adequate for your first projects. As you progress you may well find that some are too limited for your needs and you require something more advanced. However, by this time, you will be in a much clearer position to know what is worth spending money on.

Before you get started on editing your material you will need to consider the settings you use for importing material and that they match whatever particular outputs you intend for your work. In Chapter 2, we discuss the aesthetics in constructing work in more detail and this will help to inform some of the considerations in regard to editing particular projects. Most software comes with a manual and there is a wealth of written texts about editing processes for film and photography. Depending on the package you use, many offer helpful online tutorials and demonstrations that you can draw on as your skills develop.

For organisational purposes we can divide project work into three stages: Pre-production, Production and Post-production. Pre-production is the origination of your project, planning and research. This may develop from simple brainstorming of ideas to the completion of final scripts or storyboards and test shots made at specific locations. Even if you intend to work in a more spontaneous, organic way it would be useful to identify the kind of images or locations and possible themes so you don't waste days or weeks looking for something to film or photograph. It is always easier to edit work whether it be sound, film or photography if you have considered from the outset the purpose of the project and have some basic understanding of how to collect and organise the right kind of material to edit into a coherent package.

The production stage is in originating material for your project. It may involve recording interviews, sound clips, filming with or without sound or taking most of your photographs. Depending on the scope and ambition of your project this may cover days or weeks of your allocated time. A suggestion when you are first starting work is to keep the projects relatively simple and short. Although you may have ambitions to shoot a major feature film on your mobile (and this has been done) it would be advisable to begin with something less ambitious for the first project.

Post-production is the final phase of work and should take up the bulk of your allocated time. This includes editing work, re-shooting photographs or film footage and re-recording sound. It is also where you will consider the addition of music and fine tune the project to a level where you feel happy to share it with an audience. There is a tendency with some students to rush this stage either in an eagerness to complete work or because time hasn't been adequately managed to meet a deadline. If you are aiming to move beyond the amateur market or even if you just want to produce the best possible work for yourself, it is important to take time to edit. It is in the editing that you try things out and through the process of doing, you discover what works and what doesn't. You might even 'test' the work by showing it to a focus group or getting feedback from friends. The juxtaposition of different elements of a project in the edit stage may offer up surprises and reveal how new and exciting meanings emerge from the way you construct the material. In order to edit work successfully you need to allow time for a period of reflection. This involves considering what you have done and asking yourself a series of questions; primarily, does the work say what you want it to say? It is through this process of reflection that you are able to create new knowledge and make progress. After the work has been completed you will need to consider how it will be distributed or exhibited and we examine this further in the final section of Part One.

## 1.6 Brainstorming Ideas

'There are many ways of defining creativity but the one I like best is "the expression of self". It captures a belief that we're all creative — though naturally some are better at it than others.' (Hegarty, 2014: 11)

John Hegarty the founding partner of advertising agency Bartle, Bogle, Hegarty (BBH) discusses in his book *Hegarty on Creativity* (2014) what inspires, drives and sustains the creative process. The subtitle of his book 'There are no rules' suggests there are no easy formulae to follow that will produce the perfect creative master-piece. This book is titled *Creative Mobile Media: A Complete Course* and has at its heart a belief that what you produce on your mobile phone matters to you and that it should have both meaning for the producer of the work and if you are to be successful, the recipient too. We can examine this further and consider what it means to produce work that is 'an expression of the self'. This is not to say for work to be creative it has to be autobiographical but that it involves you feeling passionate about the subject matter you choose; that you are committed and prepared to invest in something you believe in. It goes back to the very first question we posed at the start: What do you want to say and why? Themes and ideas can be found throughout this book and learning about how others have approached the researching of a project can be useful. However, the very best ideas and themes are likely to come from following your own interests and what you want to produce. If you go on to work in the creative industries you may be working to a strict brief where the client has laid down the criteria for you to follow. But, if we consider the approach Hegarty suggests, then even within such a rigid structure there are ways we can connect with and develop our own ways of relating to the project that offer something of ourselves and so something unique.

A brainstorming session is one of the simplest ways to stop over-thinking a topic and get you to focus on what matters to you. The easiest way to start this is by using an old fashioned pencil and paper and write down whatever comes in to your head. What interests you? What do you like doing? What do you like watching on TV? What makes you happy? What makes you angry? Do you belong to any clubs, or interest groups? Have you any older family members whose experiences you might want to record? By putting yourself at the centre of this initial brainstorming session you can start to prioritise any ideas or themes that standout or have some special meaning for you. From your initial list you might want to do a further round of mapping. This time, putting in the centre of your page the strongest idea or theme from your first session. You can now look at ways of connecting new ideas and themes that spring to mind from your central idea. As with the first exercise don't over-think this. Write down whatever comes to mind, however strange or silly this may sound.

It is worth returning to the medium you are hoping to work in, still photography, moving image or sound and consider what is it about the mobile phone that you can utilise to support the development of your ideas. Again you could use brainstorming to help you link your ideas to the technology. Some of the most successful student films from my classes have explored mobile phone culture itself using a mobile phone. For example, a group of students made a micro drama about the frustration of a boy trying to have a relationship with his girlfriend who was constantly texting her friends. Another successful micro drama created a fictionalised account of how a drunken date was recorded on a mobile phone and broadcast globally via social media sites. The work was created for use in high schools to raise awareness of the dangers of excessive use of alcohol, *sexting*, personal safety and privacy.

By engaging with some of the information presented in this chapter you will develop a skills-set that will enable you to do some of the basics in both stills photography and moving image. At the very least you should be able to create some still images and some film footage and upload these to editing software on a computer where you can start to make informed selections from all the material you have collected. You will also have started to brainstorm ideas and think further about the kind of themes and topics that interest you. This is still a long way from creating work for a purpose and understanding the changing aesthetics of mobile culture. You may re-consider at this point your aims and rationale in relation to what studies have shown to be common usage and how your experimental work is shaped by or differs from this. You might want to pause from the practical work and read one of the chapters in Part Two of this book. In Chapters 5 and 6, I outline approaches and research methods behind two of the mobile films I made from the point of view of a practitioner/theorist. If you are keen to develop the practical work then continue to the next chapter which gives some ideas for further projects after a critical look at aesthetics.

# Chapter 2

# Mobile Aesthetics and Criticism

*Getting Started* introduced you to the first things to consider when developing projects that use the functions of your mobile phone. This chapter builds on those suggestions through a discussion of 'outputs' with a critical focus on the aesthetics of creative photography, moving image work, sound and the archive.

Aesthetics derives from the ancient Greeks branch of philosophy concerned with notions of beauty and questions around how one defines what is beautiful. If, for example, an object or painting is considered to have a pleasing appearance, how is that judgement arrived at? What makes it 'art' and can anyone create art? Many of the concepts emerging from the study of aesthetics have, since the 18th century, shaped current thinking as to how visual art is considered. The language used to discuss art still resound with terms such as 'sublime' and 'beautiful' to describe an aesthetic experience (Slater, 2015). In considering what aesthetics means in the 21st century, questions that attempt to define art have no less significance than those from previous eras. Notions of judgement, value, taste, originality, authenticity, emotion, expression and intention are debated and contested on the form and content of a work of art. Whilst some art may communicate ideas to a wide audience, meanings may not always be transparent or universally understood. In fact, forms of visual communication are deeply affected by cultural conventions. For example, our sense of space, balance and pattern are thought to be affected by the direction of our writing; from left to right, right to left or top to bottom (Kress and Leeuwen, 2009). Western culture dominates much of the writing on aesthetics but it is worth drawing attention to the long traditions that flourish in non-western art and how this has contributed and influenced some of the rich and diverse practices we see across the world. An example of this can be seen in the second half of the 19th and early 20th century when Japanese culture became an inspiration in Western painting and photography (Jacobi and Kingsley, 2016).

In this book, I have often conflated the terms 'media work', 'project' and 'artwork' to encompass the creative uses of mobile media rather indiscriminately. However, in a discussion of the creative practice of mobile media it is important to consider what is meant by creativity and how that, in turn, is shaped by the histories of the study of aesthetics. Students may be using their phones to create 'works of art' intended specifically for the gallery space and be familiar with some of the institutional practices that have evolved and which both define and limit notions of what is considered art. On the other hand, if you are making a documentary film or creating documentary photographs the balance between aesthetic considerations, intentions and content are even more complex and nuanced. Many of the new mobile creative practices merge, sample or hybridise particular forms and have prompted scholars to assess whether a new aesthetic language may be needed in order to articulate these outputs (see Chapter 8). We will consider approaches to engaging with work that allows consideration of some of these issues and fosters a critical examination of aesthetics and ideas within the context of production and distribution. We will examine popular practices across moving image projects, photography and sound and how these may be extended through the specificities of making creative mobile media.

## 2.1  Media Criticism

There are key questions we can ask about any media or art that helps in understanding the purpose of the work. Terry Barrett in *Criticizing Photographs* (2006) details how reading about and critiquing photography (and I would add, all media) enables one to appreciate the work more fully. This is not about detracting from the pleasure or enjoyment of looking at and experiencing the film or the photograph or listening to a soundscape but about extending knowledge and having a tool kit that enables us to develop an enquiring mind. Barrett uses the term *criticism* to represent a complex set of activities and meanings rather than a simplistic set of negative connotations. He cites Morris Weitz (1964), an aesthetician and literary critic, on Shakespeare's Hamlet as an influence to his approach. Critics, according to Weitz, do one or more of four things: describe, interpret, evaluate and/or theorise (Weitz in Barrett, 2006). These categories can be used in a practical way to raise questions about an artefact. The process of critique may draw attention to areas where knowledge is lacking and suggest further research to extend our understanding, including acquiring an historical overview of particular practices and contexts of production. The processes of criticism Barrett is alerting us to move beyond aesthetic judgements and focus on the questions and issues that are often raised in the analysis of art. This process can be applied to thinking about media projects in general and learning about how other producers contextualise and articulate their practice. In turn it can be applied to the production of your own work in relation to mobile media and be a useful staring point both in and outside the classroom.

Turn to the photograph titled: *By the Sea* Fig. 2.1 and using the categories of description, interpretation and evaluation think how you might start to critique the image. Consider how you would do this as, (a) part of a discussion and (b) in a written form. Think about the kind of language you might employ and the differences between a written account and the spoken word. Some of these considerations will be expanded in the chapter on exhibition and promotion and should be helpful in writing synopses and publicity material.

## (a) *Description*

When you describe something you are trying to put into words how something looks. It is a way of communicating to others in such a way that they can picture the physical details in their mind. You might

Fig. 2.1   By the Sea.

include what you can see within the frame of the photograph; the setting and sense of place. Are there people within the image and if so what are they doing, what do they look like, what clothes are they dressed in? You might continue with the description by including colours, if there are any, or state that the photograph is monochrome. Descriptions need not be long; the more concise the better, whether spoken or in a written form.

## (b)  *Interpretation*

To interpret is to consider what something means. It is closely related to description and some would argue that the two cannot be easily separated. When you describe you are, at the same time, interpreting what it means; you are making sense of it. We can also consider interpretation in terms of what the artwork represents. Barrett suggests that to interpret is: 'to account for all the described aspects of a photograph and to posit meaningful relationships between the aspects' (2006: 43). By examining Fig. 2.1 in more detail we can start to move beyond the description of what is in the photograph and think about what it means. Traditional art criticism would begin by considering the intention of the artist or photographer and where the work can be located in terms of its form and style. That is, does it belong to any traditions or movements? For example, is it surreal, or taken in a realist or formalist style? Can it be compared or contrasted to existing work? What was the purpose behind the creation of the work?

In the latter part of the 20th century other types of criticism emerged that downplayed any pre-existing knowledge of the artist's intentions and focused on a more *ahistorical* search for meaning through a 'reading of the text'. What metaphorically and symbolically could be interpreted that would help to decode the meaning? What emotional response does it trigger in the viewer? Interpretation now encompasses a wider approach and historical specificity, if relevant, is part of a much wider set of questions we might consider including how perceptions may be altered by our own positions or beliefs. Might an historian give a different interpretation of the image than, say, a psychoanalyst?

Media criticism has been concerned with a specific kind of terminology and you may come across accounts that discuss *form* and *content* as part of a set of aesthetics concepts that are deemed relevant to the interpretation of the object under analysis. Form, according to Barrett (2006), refers to how the subject matter is presented; it shapes the content. A discussion on form would include composition and how the elements are constructed and arranged. In photography, this would also extend to include style, texture, contrast, depth of field, focus, tonal range and use of colour. We might look at how the camera was used in terms of angle and how it positions the viewer in relation to what they see. A critique of form may also encompass how these formal elements contribute or not to a sense of scale or symmetry and balance understood as part of the 'principles of design' (2006: 27).

## (c)  *Evaluation*

An evaluation is a summative process. You consider all the evidence to support your critique, including your own stance, as you come to make a critical judgement of the work. An evaluation is never neutral but in attempting to assess a work it may be helpful to ask further questions. Is the photograph, for example, effective in communicating something to you the viewer? Does it reinforce or change the way you think about the world? Do the composition and technique add something to its meaning and value or do they distort, confuse or detract from these aspects?

## (d)  *Semiotics*

Semiotics, the study of signs, has been used as a methodology and analytical tool in exploring media texts, in particular as a way of explaining why specific things mean what they do (Long and Wall, 2009). Text, in this sense, refers to any medium such as painting, film, photography, television, advertising or print media. Ferdinand de Saussure, a linguist, was one of the key thinkers in examining how language was structured and applied the term *semiology* to the study of signs and meaning. Charles Peirce, an American philosopher who used the alternative term *semiotics*, developed Saussure's theory around signs, symbols and meaning to include the idea that some signs had iconic and indexical links to the objects they represent. This is relevant when we consider photography and the resemblance the object in the photograph has to its physical object in the world. Early photograms illustrate how objects change the way we see light and their imprint can be recorded on light sensitive paper or film. The object once existed as a physical presence and what we see in the image looks like the physical object in a way that is more realistic than, say, if depicted in a painting. Other theoretical readings on photography have questioned the one-to-one correspondence and speak more of it being a representation of an object in two-dimensional space that always needs interpreting to give it meaning. Interpretation in this sense also means an examination of meaning through usage and how this is regulated through institutional practices (see Watney, 1986 in Evans and Hall, 1999; Tagg, 1998).

Roland Barthes applied the theories of semiotics to popular culture and the mass media and expanded not only the ideas of how meanings are created but how it is possible to expose, through analysis, the underlying ideologies and cultural 'myths' that may be operating. Description and interpretation have parallels with Barthes' ideas of denotation; the literal or descriptive message, the elements in the scene and connotation, the symbolic message. How we connote meaning is a form of interpretation and this is done in relation to the symbolic codes at work within culture to signify specific things or feelings. Barthes went on to look at myth or ideology at work in popular culture. The chapter on *Rhetoric of the Image* (1964) from the book *Image Music Text* (1977) explored the symbolic meanings coded in an advertisement for Panzani pasta and tomato sauce and how *Italianness* is conveyed. The ad worked by attaching connotations of all things Italian including the symbols of market freshness, to a French manufactured product. In *Mythologies* (1972), the cover of *Paris Match* magazine (1955) was analysed revealing the symbolic and ideological meanings at play behind the image; how a more political reading of a black soldier saluting the French flag in the context of France's colonial past gets de-politicised or neutralised. If you have not come across semiotics before, a great deal of scholarly work has been written around the subject. You may find it interesting to read more widely on this by first looking at the original texts of Barthes (see Bibliography) and then more current writing on the subject including accounts that critique semiotics as a methodology.

Media criticism involves gaining knowledge about someone's work created specifically for others — an audience. In evaluating work, our judgements are called into question including any emotional response we may have to the work. We may speak about how we 'like' or 'dislike' something but it becomes more useful if we start to consider why that is so. What pre-existing biases and assumptions may be influencing our responses that we might previously have not considered? It might be useful to consider those aspects of our life, experience and culture that may influence our decision-making processes so we can not only be open to new ideas but be mindful of how feelings and opinions are shaping judgement. To critically engage with work is a two-way process and reflecting on our own role in that process is likely to lead to more informed opinions.

Cultural theorist Stuart Hall, author of the seminal chapter entitled: *Encoding/Decoding* (1980) from *Culture, Media Language;* has influenced how we think about the production, consumption and reception of media texts and the role the audience plays in the communication and circulation of meaning. Hall is cited here as his work has generated much scholarly writing, referencing and criticism and forms part of a canon of thinkers who have been influential in the 20th and 21st century on Cultural and Media Studies. You might balance Hall's perspective by looking at the work of some of his critics. For example, Kim Schroder's critique of Hall and the classic model of encoding/decoding is examined in *Making Sense of Audience Discourses: Towards a Multidimensional Model of Mass Media Reception* (2000). Reading a range of texts that critique particular theories is part of a research methodology that allows the reader to examine the evidence presented and assess the validity of any claims made.

French sociologist, Pierre Bourdieu is another influential writer in cultural theory. In his book, *Distinction*: *A Social Critique of Judgement and Taste* (1984) he argued that power in society was created and maintained through access to capital. This moved beyond economic capital and operated at the cultural and symbolic level. He used the term *habitus* to describe how cultural capital was acquired through belonging to particular social classes and was manifest in everything from taste, manners, dress and the kinds of cultural pursuits one might engage in. Aesthetic judgements are often considered a matter of taste and if we take Bourdieu's analysis we can see how some tastes are given more legitimacy and value than others. Bourgeois culture accrued more value in French society and this class maintained its position by distancing itself from working-class culture. So, for example, does one class have more cultural capital than another to appreciate a fine painting or a sculpture?

Aesthetic choices therefore become distinctions that mark a person out and set them apart from other people. We can see this operating in a variety of ways that can be class based but which, in the 21st century, can operate across a more complex, nuanced terrain including the consumer-driven culture of late modernity/postmodernity. Such is the anxiety around matters of taste that getting things 'wrong' and thus be open to ridicule, has created an industry of *taste-makers* dedicated to helping people make the 'correct' lifestyle choices on everything from what to eat, how to dress and furnish one's home to the films we might watch (Powell and Prasad, 2007). Informed by the work of Bourdieu, Beverley Skeggs extend ideas on class to consider other inequalities which operate in terms of gender and representation in her book *Formations of Class and Gender: Becoming Respectable* (2002, ed.).

In developing our skills of critical analysis, media can be analysed from a variety of perspectives that, in turn, can reflect and shape our own perspective. We haven't begun to think about psychoanalytical accounts that examine how the mind may play a role in our response to media, or made explicit the approaches of philosophical thinkers such as *Gilles Deleuze* or *Michel Foucault*. It is impossible within the scope of one book to give every possible approach but, hopefully, you will use some of the references in this chapter to further your reading and research on the subject to follow what interests you the most. By adopting a critical approach we can learn to recognise the different viewpoints and agendas that are part of academic writing; to consider what gets privileged as part of the canon, what gets left out and why. We are looking at how relationships of power operate and this in turn can be seen as a very specific approach with a history and a body of 'key' thinkers and writers attached to it.

Thus, armed with some of the methodological tools of media criticism, the following sections look at some established and evolving techniques for creating media across moving image, photography and sound and how they may be adapted to include the specificity of mobile phone technology.

## 2.2 Photography

In *Getting Started* how the camera in a smartphone operates was considered in relation to more conventional digital cameras. What make a 'good' photograph was thought to be determined by a number of key elements that are context specific. What is meant by this is that there are some exceptions to the dominant codes of aesthetic traditions and that the context of how the work was produced, in what situation and for what purpose can affect how we critically engage with the image. An image inserted into a news story may have been taken to illustrate an event as it unfolded. There are examples of this kind of imagery taken by smartphone users at the scene of newsworthy events. Speed and details are usually what is required and we can forgive the photographer if his/her compositional skills and lighting techniques fall below the conventions of the established photographer. Professional news photographers are usually experienced enough to do both but the increase in citizen journalism has seen many photographs included in print media and websites that don't follow the usual rules. They merit inclusion because they do something more than create a well-composed image (see some of the discussions in Chapter 8: *Mobile Journalism*). We can also consider any number of images classified as *fine art* photographs that are beautifully composed, balanced and where the lighting demonstrates a carefully crafted piece of work. We can admire the photograph and the skill involved and yet there is something lacking; it remains devoid of any meaningful content or is unable to generate a connection at an affective level.

We take, as a way of developing our photographic skills, some of the steps professional photographers traditionally take on their journey to producing memorable images. The purpose is to develop what professionals call a 'photographic eye'. As with most, skills-based work practice is key but being able to reflect critically on what is produced is also about gaining valuable knowledge; becoming a thinking photographer able to create meaningful content. In terms of the still image a number of elements can be assessed and together they contribute to the overall 'quality' and 'aesthetics' of the image. These include:

- lighting
- composition and balance
- focus and point of view
- tonal range/use of colour

You may want to refer back to the exercise you completed earlier and review your images in light of the above categories. It is important to remember in any discussion on aesthetics that nothing is straightforward. Opinions are made in relation to the prevailing cultural norms and taste and judgement are cultural constructs. We also need to be mindful of the fact that creative mobile practice is at the frontier, if not forging a new aesthetic, of changing the way we think about the world. That said, the chapter continues with further elaboration on the conventions that have shaped a more traditional practice as a foundation for moving beyond those constraints and boundaries.

Consider how light is utilised to add shape and depth to your images and the choices and control you have over what parts of the images are illuminated. Natural light refers to normal daylight but this can vary from minute to minute, from season to season and depends on your geographical position. For particular kinds of photography, such as landscape photography, you are reliant on the condition of the available natural light. Many smartphones carry a built-in compass and this may prove useful in helping you determine the sun's position and where the light and shade is likely to be. Direct overhead sunlight during summer months can create harsh shadows. Light levels in winter months in the UK drop very quickly after mid-day. Cloud cover diffuses light and grey skies can produce a very 'flat' looking image. It is this ability to assess lighting conditions, light sources, highlights and shadows and how you can use what light you have effectively, that improve your skills.

Not all photography projects are shot outdoors and how you use available light, including artificial lighting as opposed to natural lighting, warrants some thought. Different kinds of light produce different tonal ranges. For example, candlelight is on the spectrum of yellow-red wavelength. Some fluorescent indoor lights produce a green tinge to images that may need correction in post-production (Hirsch, 2012b). It is important to understand how different light sources affect your images as they can be utilised for creative effect or balance using image manipulation software such as colour correction filters.

We also respond, as humans, differently to different tonal ranges and colours and this may be worth considering when thinking about what kind of emotional response we want from the viewer (Wells, 1910; Kurt and Osueke, 2014). In post-production, black and white and sepia filters have become commonplace to give an aged look to an image. Sepia in particular can signify nostalgia but the repetition of such photographic codes have led to its ubiquity so that meanings become clichéd. Considering the whole colour palette, however, can add a further dimension to creating an emotional engagement with your viewer. Used subtly, warm tones — reds, chocolates, oranges — give a feeling of emotional warmth. Red, for example, has been associated with arousal of the senses. On the other hand, colours on the blue spectrum are described as cold as these can create a feeling of emotional distance (Kurt and Osueke, 2014). If you were asked to create an image in the style of a *gritty, urban noir* think what colours and tones would be most effective and how you might also use light and shade. Colour, light and shade, soft or hard shadows can all have a psychological effect on the viewer and be mood altering. This is not just about the biological attributes of perception but also how our interpretations can be culturally determined; cultural patterns can shape emotional response to colour. For example, in Asia orange is associated with spiritual enlightenment and has positive attributes. In Europe and North America, it is associated with hazard lights and warning signs (*ibid.*).

Some of the images you have produced on your phone may be landscapes of places visited or portraits of friends and family and hopefully they will have meaning for you. Personal photographs can be a starting place to consider how you want to develop your work. Many photographers and artists have turned to autobiography to express their ideas and have also managed to create something that appeals to a wider audience. The size of the camera phone makes it uniquely placed to be unobtrusive and discrete. Furthermore, you are likely to have it with you most, if not all, of the time making documenting aspects of your life an easier process. Deciding to share autobiographical work with others and make it into a public-facing project needs more critical thinking. Working in autobiography raises ethical consideration for family members, friends and your own personal well-being. You may wish to read the chapters on *Questions of Ethics* and *May Days* as well as researching other practitioners who use an ethnographic/autobiographical approach before continuing. For further reading consider Annette Kuhn's (1995) *Family Secrets: Acts of Memory and Imagination*, or Marianne Hirsch's (2012a) *Family Frames: Photography; Narrative and Postmemory*.

Working autobiographically isn't for everyone and the following sections focus on aspects of photographic practice that are both suited to mobile phone technology and draw on history to contextualise and locate the emerging aesthetics.

### 2.2.1 *The New Flâneur, Flâneuse and Street Photography*

Charles Baudelaire first described the *flâneur* as a character who inhabited the streets of Paris, in his 1863 essay *The Painter of Modern Life*. Strolling along the pavements and boulevards and observing the fleeting, transient moments of modern life, he captured the spirit of the times. Poetry, painting and writing of the period documented modern urban life through a *sense of feeling* juxtaposing both the wonder and spectacle of the new with the fragility and anxiety generated as part of the conditions of modernity.

Walter Benjamin, philosopher and cultural critic, documented transformations in the city of Paris in the *Arcades Project* (1927–1940). He was interested in the micro details of a city and set out to map the changes in retail culture, and through this, the shifts in capitalism and the birth of modern consumerism. The *flâneur*, for Benjamin, was not just a character but an archetype of the urban experience: "a botanist of the pavement" with *flânerie*; both a practice and a methodology to explore the modern city (Benjamin, 2002). He asks us to consider the question: How can we know a city and how can we represent it? The methodology of the *flâneur*, becomes more than documentation; it can be part of a political act, a tool to expose power relationships at work in the transformation of any city. Benjamin considered those who were alienated as a result of capitalism and how they could be reconnected with the city. The demise of the 19th century *flâneur*, has been replaced in the 20th and 21st century by a more contemporary notion of *flânerie* and artists, photographers and writers who explore the tensions of an urban environment through documentation and exposè.

If we consider a global city like London, multiple accounts of the city are to be found ranging from historical documentations to Peter Ackroyd's personification in *London the Biography* (2000). We have more 'official' representations that present London in terms of capital, commerce and tourism. Barthes and semiotic theory highlighted how particular versions of events are created and legitimised as official versions. It might be worth bearing this process in mind when we begin to think about cities, as Benjamin did, when critiquing bourgeois 19th century Paris in the Arcades Project. There are many representations of a city but we can start to understand more when we move from the 'official' versions to the micro levels of exploration as well as considering how the city is inhabited. The methodology of the *flâneur*, therefore, gives us a particular technique to observe and engage with the city. The *flâneur* was a solitary character who took up positions on street corners, strolled through the commercial districts or sat at a pavement café to observe, discretely, city life. He was the anonymous face in the crowd. His viewpoint was distanced and detached from the subject matter but what he saw could promote reverie and lend itself to narratives of what if…? We might consider the technique as a snapshot of a moment in time, capturing surface features rather than depth. But in the micro observations we can start to uncover the layers that make up a city. We can observe the parade of passers-by, note their style of dress, the pose, the manners of street life, the chance meeting, the stranger in the crowd. Twenty-first century *flânerie* is not confined to the male, middle-class, figure that was romanticised in 19th century Paris. The *flâneuse* is increasingly occurring in writings that map the use of public spaces from a female perspective. Lauren Elkin notes in *Flâneuse: Women Walk the City*; 'It would be nice, ideal even, if we didn't have to subdivide by gender-male walkers, female walkers, *flâneurs* and *flâneuses* — but these narratives of walking repeatedly leave out a women's experience' (2016: 20).

The smartphone can make everyone a camera-*phoneur* but the method still calls for the eye of a detective or sleuth. It also requires some consideration regarding the ethics in employing methods of detachment and discreet observation and when ordinary people may become the subject matter. It might be considered intrusive and in some cases may raise legal as well as ethical questions. If your subjects are recognisable in the image you should get their permission and if work is to be published this needs to be in a written form. Such a challenge may lead to unexpected and creative ways of photographing street-life and photographers have employed the use of reflections and low-angle camera positions as a way of capturing movement and imagery that doesn't involve recognisable subjects (see Fig. 2.2).

The Paris Arcades documented by Benjamin also became the haunts of the surrealist artists. Two of its key members, Andre Breton and Louis Aragorn, produced what Merlin Coverley has described as

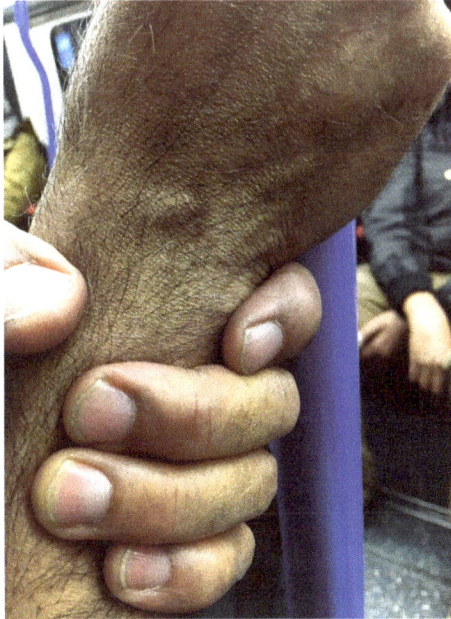

Fig. 2.2   On the Piccadilly line.

one of the first *psychogeographical* novels. Lacking any plot, the writing was from a perspective of the inner desires of the characters and aimless strolling through the streets of Paris. Automatism and the unconscious were part of the "free-floating" exploration of the city and resulted in writing that was shaped by "coincidence" and "uncanny juxtapositions" (2010: 74).

Guy Debord, as part of the *Situationist International* art movement, was to develop some of these ideas into the *dérive,* which literally means drift. 'In the *dérive* one or more persons during a certain period drop their usual motives for movement and action, their relations, their work and leisure activities, and let themselves be drawn by the attractions of the terrain and the encounters they find there' (1996: 22). Unlike the surrealist artists the *dérive* wasn't so much driven by links to unconscious processes of the walker but the geographical features and spatial configurations of the city. On a *dérive* the participants needed to be conscious of the psychogeographical relief of the city, the feelings evoked and the "fixed points" and "vortexes" that could block free movement (*ibid.*). Surveying the city through the technique of the *dérive* allowed, according to Debord, for psychogeographical articulations and plotting of the boundaries, junctions, barriers and borders that are not just physical but are part of the 'atmosphere' of a modern city and would give a new and 'authentic' experience to urban life.

Psychogeography is at the heart of a number of literary works most notably the writings of Iain Sinclair. *Lights Out for the Territory* (1997) uncovered the "secret histories" of London through a series of walks that traversed the city. History, power and money underpinned many of the reflections linking the geography and people of London. In *London Orbital* (2002), Sinclair set out to walk the M25, keeping within the "acoustic footprint" of the motorway that encircles the city and excavating the layers of social history that surround the capital. *Hackney, That Red-Rose Empire* (2009) took an even greater political turn in the part documentary, part fictional account of the London borough as it prepared for the 2012 London Olympic games.

The methodology of the *flâneur,* the *dérive* or drift photography and psychogeography have been adopted as approaches to explore the urban environment. They contribute to the repositioning of cities that provided contrasting and conflicting juxtapositions and illuminate what shape the geographical, psychic and cultural landscape, including how notions of power, economics, class, gender and race may operate. The camera, by the end of the 19th and early 20th century, had become the perfect recording tool for documenting the changing landscapes of the city. The genre of *Street Photography* emerged as a distinct set of practices that can be defined in its loosest form as: 'any photograph made anywhere in a public place' to 'un-posed scenes that trigger an immediate emotional response' (Seaborne and Sparham, 2011: 7). By extension, the mobile phone camera is the perfect tool to record and document street life.

The following projects are suggested as a starting point for the exploration of your nearest town or city:

---

**Photography Projects**

1. Become a *flâneur/flâneuse* and stroll around your nearest city as the 'detached observer'. Use your phone to document street life. Use the memo function to add a commentary on what you see, hear and feel. Remember to keep your work legal and ethical.
2. Go on a *dérive*.
3. Walk a linear route from two fixed positions and document your journey at fixed intervals. Use the images as a starting point to research the history of the area or create a narrative.
4. Consider the entrance and exit points of a city for a project. Remember you will need permission for photographing on private property including railway stations.
5. Ghost Signs — Capture the history of the buildings by photographing old shops signs and advertisements. Remember to look up!
6. Borders and Boundaries — Explore the geographical and the psychological borders and boundaries within villages, towns or cities where you live.

---

(*Continued*)

*(Continued)*

---

**Sharing your work with others**

Edit the images into a 2 minutes slide show. Add sound/music using a copyright free source or make your own soundtrack. Upload the completed project to a themed web page to share with an online community.
Use the images to illustrate a blog.
Make an e-book or printed photographic book.
Use a pop-up gallery space to display your project.

---

## 2.3 The Moving Image

The term moving image relates to all lens-based practices that include the use of moving visuals that are time-based. This may include traditional narrative filmmaking made for the cinema or television or more experimental gallery-based or web-based projects. You might want to start thinking about the moving image and what kind of practice interests you. Consider films in any genre that you find appealing. How do they engage with an audience? A successful blockbuster may fail to have a convincing narrative and plot but nevertheless allows us the pleasure of watching the spectacle, the scale and special effects that appeal to the senses. We may encounter a moving image in a gallery installation that puzzles us but leaves us wanting to know more. A film may open up a new perspective on how we view the world.

The criteria you engaged in for critiquing work at the start of the chapter might be useful to apply to films you have enjoyed and those you feel are less successful. Start to analyse what, for you, are the important qualities every film should have and then consider what it is you want to say and how you want to say it. These seem like very basic questions but it is surprising how many students start from a position of what they think they ought to do rather than something they care about enough to sustain the research and production across a period of time. You may be lucky enough to be affiliated to a group, club, society or organisation that has asked you to produce a film about them. You might just want to try your hand at making a micro-horror film because that is something you love watching. You will also find in Part Two of this book, accounts of how hybrid mobile documentary has been utilised to explore specific ideas and create stories. The quality of smartphone cameras mean the limits for the kind of films you can make are much reduced. *Tangerine* (USA, 2015) co-written and directed by Sean Baker received publicity as the first smartphone feature film to have a wide theatrical release across the United States. It was shot exclusively on an Apple iPhone 5s with an anamorphic adapter for wide screen.

### 2.3.1  *The Narrative Film: Telling a Story*

Whether you set out to create a work of fiction or to produce a documentary you need to consider the narrative. What story do you want to tell? Whether factual or fiction, you need to have engaging characters and dramatic tension. The best stories say something about the human condition and have something the audience can relate to. Writer Peter Brooks introduced his book *Reading for the Plot* with the following words: 'Our lives are ceaselessly intertwined with narrative, with the stories that we tell and hear told, those we dream or imagine or would like to tell, all of which are reworked in that story of our own lives...' (1984: 3). In an examination of what it is that shapes stories and gives them meaning, Brooks discusses the centrality of narrative as to how we understand and make sense of our own lives; how we use narrative and plot to negotiate versions of reality. In this sense, creating order and meaning through a narrative structure of a film has similarities to how we use narrative technique in everyday life; to provide meaning and order to the randomness of the human condition. Exploring narrative can provide useful tools in thinking about film structure and how the stories we tell might communicate and have resonance for others.

Vladimir Propp a Russian formalist studied hundreds of Russian folk tales and came to the conclusion they could all be reduced to specific key elements in terms of their structure. These elements were the building blocks on which all stories were based and could be broken down into a number of different characters or archetypes and the functions they performed. He outlined 31 functions, understood as actions, the characters performed and the consequences of such actions. Some of the popular characters outlined in *Morphology of the Folk Tale* (1927) we can recognise from popular folk tales; the hero and villain, the dispatcher, the helper, the false hero and the princess. The hero is introduced into the story and is either dispatched on a quest or journey or has to overcome some hardship or difficulty. He is tested on the journey but gets help, sometimes in the form of a magic potion from a helper. Along the way, he meets a villain who tries to prevent him from reaching his goal and who has to be defeated in order for the hero to succeed. The villain gets punished or killed and the hero is rewarded by marriage to the princess. Not all characters need appear in every tale nor all 31 functions need be present but, Propp argued, the sequence of events within the tales remains the same. A Proppean analysis of dramatic structure in narrative action has been applied to many different kinds of stories, including non-fiction (Machill *et al.*, 2015). Propp's writings have influenced how we might think about characters and their actions and can be updated to more realistic settings in a modern world. For example, the hero is not exclusively male but defined as the main character driving the action and who, in more simplistic stories, is morally 'good' (see Chapter 9) and the magic potion can be replaced by a particular skill that is acquired with help from a friend or supporter.

Based on Vladimir Propp's analysis of the folk tale, Tzvetan Todorov divided narrative into five key stages that govern the internal workings of the story world. These are: (1) Equilibrium (2) Disruption (3) Recognition (4) Resolution (5) New Equilibrium (1977). The narrative begins from a position of equilibrium or balance inferred in the story world. Stage two introduces a disruption that creates a state of disequilibrium. The disruption can take many forms and may be as dramatic as a bomb exploding or a death, or subtler such as a chance remark around the family dining table triggering a series of events or actions. It is the cause and effect of the disruption that propels the action and the narrative forward. Once the disruption has been recognised and dealt with, equilibrium can be re-established. The world may be a very different place from the equilibrium of stage one but some sort of resolution will have taken place to restore a new balance. These stages have been documented as the 'classic narrative tradition' forming a common structure for much of the commercial films produced in North America and Europe that have dominated mainstream cinema. That is, they have recognisable schema based around conflict and resolution and through the repetition of particular codes and conventions that become familiar to us, create a shared meaning (Long and Wall, 2009). Propp and Todorov's theories on narrative structure can be applied to drama and non-fiction and you may want to analyse some popular films in each category, working through their structures and characterisation. Who might be positioned as the heroes or villains in a current news story or in a wildlife documentary for example? Are there any films that don't conform to such a structural analysis?

Of course, it may be that your rationale for creating a mobile film is to explore other ways of telling a story and in opposition to what might be regarded as a form of cultural domination by the US and Europe of cinema. Exploring narrative film outside of North American and European traditions would be a starting point for analysis between any similarities and differences. You might also consider ideas around the *Transcendental Narrative* outlined by Erik Knudsen in (De Jong *et al.*, 2012) *Creative Documentary* that approaches storytelling by exploring narrative outside the classic traditions and seeks to engage with "participatory or spiritual feelings" (2012: 132). It is a form of storytelling that attempts to immerse the audience in the imagery or the events rather than following the action. Knudsen suggests there is still a reason for the story and the ending will still seek change or revelation, but the approach is different to the classic narrative (*ibid.*).

Using narrative as a methodology to convey ideas is a well-trodden path in creating films that can relate to a wide audience. Rabiger, suggests: 'Every compelling story, fictional or documentary, has characters striving to accomplish something and overcoming obstacles from their circumstances' (2009: 12). He goes on to suggest the characters don't have to be people and that some of the most enthralling stories can document, for example, a beetle climbing a stalk in order to take flight (2009). This sets up part of the drama in the narrative but gives the audience something they can identify with at a deeper level. It may only be a beetle but we invest in the struggle unfolding on the screen as it overcomes the obstacle, climbs the stalk and it and we are both rewarded if the creature succeeds in taking flight. Creating situations where audiences can identify and empathise with some of the characters has been a key function of narrative in getting us to engage with the story and keep us watching.

## 2.3.2  *Documentary and The 'New' Realism*

One of the most contentious subjects, in media production is the relationship between aesthetics and documentary practice, that is, the relationship between how something looks and its purpose. John Grierson is often referred to as the founding father of documentary filmmaking. He defined the practice in the 1930s as: "the creative treatment of actuality" (Grierson in De Jong *et al.*, 2012: 19). This suggested a relationship between the creative processes of editing and the actual events recorded. It highlights the fact that from the outset documentary is a construction rather than a totally objective medium. There are now numerous interpretations of the many documentary practices that have emerged and the relationship filmmakers have to documenting real events. Bill Nichols, in *Introduction to Documentary* (2010, ed.), defined four main categories of practice that included the *Observational* mode, the *Expository* mode, the *Interactive* mode and the *Reflexive* mode. These were not exclusive categories and some of the styles might overlap within a single film. De Jong *et al.* in *Creative Documentary* (2012) consider an even more eclectic mix including the *poetic* mode and hybrid styles that are part of the contemporary practice of the 'total' filmmaker. By this they are referring to those engaging with the research, making, editing and distribution of work and are likely to be involved with non-traditional platforms such as the Internet (*ibid.*).

It can be argued that documentary filmmaking has no single, uniform practice and its form — the way the story is told in terms of style, structure and techniques — is multifarious. Rabiger also adds that the modern documentary 'avoids telling us what to feel and think' (2009: 16). The viewer, through this mode of storytelling, is exposed to contradictory accounts of events that allow the evidence to be challenged and seen from different viewpoints. However, there are particular characteristics associated with the practices that allow us to distinguish documentary from works of fiction. Firstly, there is the relationship to actuality; to events that have taken place or are taking place and may use reconstruction of such 'real events' as a technique of storytelling. Secondly, it is distinguished from other non-fiction forms by the underpinning of a range of values, choices and consequences. Modern documentaries are likely to be 'socially critical' (Rabiger, 2009).

What these examples demonstrate is that representing actuality, real events or working with ordinary people (the untrained actor) continue to be problematic but nevertheless are still bound by particular codes, convention and expectations in how filmmakers construct and represent 'real' life. These may shift and change over time and this can be demonstrated by engaging with some of the films listed at the end of this chapter that range from the 1920s to present day. Audiences accept the dominant codes and rely on the validity and credentials of the filmmaker to represent events in a fair and honest way. They can be held to account where expectations are not met, particularly when the audience feels they have been tricked or deceived (see Chapter 9 on ethics).

A second burden rests on the documentary maker not just in terms the kinds of reality and content they depict but *how* that is done in relation to the subject and matters of aesthetics. Photographer, Sebastiao Salgado, came under particular public scrutiny when Ingrid Sischy accused him, in her article *Good Intentions*, of lacking sensitivity by paying more attention to aesthetics than to the suffering of the people in his photographs. 'His compositions, crops, lighting, angles and toning stand in sharp contrast to the usual lack of insistent style in photojournalism' (1991: 90). We can unpick this a little further in terms of the value-judgements at play. In serious subject matters such as war reporting, depicting poverty or trauma, creating a 'beautiful' looking film or photograph — paying attention to the common notions of aesthetics — should, it is argued, be secondary to the message being conveyed. If we agree that the message or meaning should be paramount in documentary then the *style* should not detract from what is being conveyed. Twenty years later, photojournalist Damon Winter was defending his aesthetic choices for taking a series of images in Iraq for the *New York Times* using the photo app Instagram. In *Through My Eye; Not Hipstamatic's* he argued: 'We are being naïve if we think aesthetics do not play a role in the way photojournalists tell a story' (Winter online, 2011). He goes on to justify his approach by stating:

> 'The problem people have with an app, I believe is that a computer program is imposing the parameters, not the photographer. But I don't see how this is so terribly different from choosing a camera (like a Holga) or a film type or a processing method that has a unique but consistent and predictable outcome or cross-processing or using a color balance not intended for the lighting conditions (tungsten in daylight or daylight in fluorescent, using the cloudy setting to warm up a scene)' (*ibid.*).

What these examples illustrate is just how contentious matters of aesthetics, balancing the form with the content, can be.

As far as creative mobile media is concerned there are many aspects of documentary practice that are well suited to the technology and it is possible some of the first films you create will come under this umbrella. As discussed previously the size and availability of the camera phone are useful for 'bearing witness' to events as they unfold. They have become an integral part of newsgathering with large institutions such as the BBC offering training in the use of these devices for their reporters. Marc Settle explains why the BBC Academy are offering such training and comments: 'They are truly multimedia devices: they enable journalists to produce, with just one machine, content for which they would previously have needed half a dozen or so' (Settle, 2015: online). Furthermore, if we think in terms of the kinds of aesthetics emerging from camera phones including shaky footage, it is fast becoming the go to 'code' to signal actuality, authenticity and 'real' life (see Schleser, Esparza and Kapur in Chapter 8 for more on mobile journalism). Established filmmakers Ridley Scott and Kevin Macdonald created the first feature length documentary film using user-generated content largely shot on mobile devices. *Life in a Day* (2011) (https://www.youtube.com/user/lifeinaday) which was described as a "global experiment" and "snap shot of the world", was edited from content shot in a single day on July 24, 2010 (Macdonald, 2011). The format proved popular and was followed by *Japan in a Day* (2012), *Christmas in a Day* (2013) and *Italy in a Day* (2014). The latest release from Google is *India in a Day* (2016) directed by Richie Mehta and produced by Ridley Scott and Anurag Kashyap. These are all fine examples of how the smartphone is changing the documentary landscape and making participatory documentary, by way of the Internet, a new and powerful medium.

To end this section on aesthetics and the moving image we turn to some rather old traditions and how these are being reconceptualised in the digital age for the smartphone filmmakers.

## (a) *Direct Cinema and Cinema Vérité*

Some of the 'new' aesthetics of mobile filmmaking aren't so new at all. Observational cinema, fly-on-the-wall techniques popular in the 1960s and 1970s, were once thought to be the purest forms of documentary practice giving unmediated accounts of actuality. This has been challenged on two counts. Rabiger states: 'An observational film makes us feel like privileged observers, but we are seldom seeing life unmediated as *transparent* film suggests' (2009: 86). Subjects know they are being watched and adapt their behaviour accordingly. A way round this is to use hidden cameras but this raises ethical issues that many contemporary filmmakers feel uncomfortable with. *Cinema Vérité* or *Kino-Pravda* uses an alternative approach. Invented by the ethnographer Jean Rouch and taking inspiration from Dziga Vertov's

(1929) *Man with a Movie Camera*, it foregrounds the importance of the filmmaker's presence and filmmaking as a collaborative act. Directors could 'probe for the truth rather than simply wait for its appearance' (2009: 88). Observational film and *Cinema Vérité* techniques were made possible through advances in technology and the production of smaller, more portable cameras. This has evolved to such an extent, it is now possible to keep a recording device in a pocket. Hand-held camera techniques, the relationship of the camera phone to the body, the use of selfie stick that incorporate the body of the filmmaker have emerged from the practice of self-filming and build on some of the philosophy underpinning *Cinema Vérité*. These techniques have become commonplace but interestingly are valued as part of an aesthetic that signals authenticity and openness.

## (b)  *Experimental Film*

There is a rich history of what can be loosely defined as experimental film and avant-garde cinema that had their origins in Europe from the 1920s. Under this umbrella are any films that don't depend on narrative techniques and include a wide range of practices. They are linked to artistic practice and are generally more at home in the gallery space rather than a cinema. A low budget is a characteristic of most experimental work. These films may be self-financed or made with small art grants and funding sources. They are most likely to have been made by an individual undertaking all the aspects of filming and editing or by a small crew. Co-operative movements, such as the London Film Co-op founded in 1966 — there are equivalents in other major world cities — not only provided both physical spaces where experimental filmmakers could share ideas, skills and have screenings but also acted as agencies and distribution platforms. Now incorporated into *Lux,* the international arts agency for the support and promotion of moving image practice, the current online space for many experimental films states its remit in the following way:

  'The particular focus of LUX is visual arts-based moving image work, a definition which includes experimental film, video art, installation art, performance art, personal documentary, essay films and animation and is inclusive both in terms of context and critical discourse' (www.luxonline.org.uk/histories/1960-1969/london_film-makers_co-op.html, accessed April 2015).

The Internet has provided a platform for the continuing practice of experimental films so keeping the spirit of this kind of work alive. A 'dog's eye view' of the world, created by attaching a mobile phone to a dog collar, created by a group of students on my own course, was a novel interpretation of the experimental film genre. Referenced in chapter one was filmmaker Clio Barnard short film *Dark Glass* (2004) created for the *Single Shot* Festival and website. For her 'single take' film she created an exploration on hypnosis, time and memory using a technique that included taping her camera phone to a hockey stick (2011, *Pocket Cinema*, BBC Radio 4, UK). There are numerous resources and online sites that cater for experimental films and the following links may be useful.

- *Short of the Week*, available at https://www.shortoftheweek.com/channels/experimental/ (accessed March 2016).
- *Experimental Cinema*, available at http://expcinema.org/site/en (accessed March 2016).

## (c)  *The Micro Drama/Micro Film*

Documentary and experimental films are genres that have adapted well to developments in mobile phone technology. More surprisingly perhaps is the rapid development and spread of the micro drama or micro film. These are tightly contained narratives with a duration of between 90 seconds and 5 minutes. They are usually shot on smartphones or made specifically to be viewed on mobile phones and tablets via Internet sites. Originally developed by amateur filmmakers with very small budgets, now both amateur and established directors are using the form to experiment with storytelling. Advertisers were also quick to see their potential and now utilise the micro film across a range of platforms to sell products or act as sponsors for events. The Pocket Cinema Film Festivals in France and Japan, which showcase micro films, were amongst the first of their kind and similar festivals in other countries soon followed. The iPhone Film Festival was launched in 2010 as an online platform (http://www.iphoneff.com/). China held its first International Micro Film Festival in 2012 and with a population of more than 1.3 billion people and an increase in growth of the smartphone, it is one of the largest markets for this new genre. More recently Pakistan launched its first Pocket Cinema Film festival in September 2015. There are now many dedicated website which operate as platforms to for screening work and most have categories for the micro drama or micro film.

## (d)  *Convergence and Hybridisation*

As we have seen the Internet provides a platform for all kinds of media, from the traditional film and documentary to more hybridised and experimental forms. The vlog (video blog) has become ubiquitous on the Internet with popular sites such as YouTube and Vimeo acting as hosts. The blog that combines informal styles of writing, graphics, photographs and videos inserts is another way for an individual to gather a following and get their interests and ideas shared with other, like-minded people. With appropriate content and publicity it has also become a source of income for the new, web-based entrepreneur. Getting sponsors and embedded links to particular products and services that target consumers has been the key to the success of many such entrepreneurs. Some of the most commercially successful blogs are lifestyle based with cooking, beauty and fashion advice mixed with more personal chatter about friends and family. Spin-offs include the production of cook books, clothing ranges and jewellery with the originators, or 'stars', becoming skilled at promoting their ideas through other social media channels such as Twitter, Tumblr, Pinterest and Instagram alongside print magazines, television and promotional tours. Two such examples are the *Deliciously Ella* blog (http://deliciouslyella.com/), which started as a food blog by Ella Woodward and *Zoella* (https://www.zoella.co.uk/), originated by Zoe Suggs when she was a teenager. *Deliciously Ella* has been running for over four years, has a large following and a cookbook on the international bestseller list. *Zoella* has over 10 million subscribers and is popular with the female teen and pre-teen age group. Suggs has launched the *Zoella* branded make-up range of products and published her first novel *Girl Online* (2014).

Simple blogs using video inserts created by using the *flip* setting on a smartphone have an 'inward-looking aesthetic' focused on personal stories and are usually filmed in the blogger's home. Affiliated marketing, sponsored posts and product placements have seen some of these become part of a multi-million pound industry. Most bloggers, especially women, use blogging to supplement other income streams or as a way of earning a little cash whilst they are at home with young children (Beugge, 2014). However, established companies are only likely to promote their products by sponsoring blogs that can demonstrate a large following. Over time the changing aesthetics, 'the look' of such websites, tend to shift to a more slick 'outward-facing' professional style that is testament to the successful links between commerce and the personal. There is always a balance to be struck as far as aesthetics are concerned. If the websites cease to maintain the personal connection that many of their followers are attracted to they lose their appeal. There are, of course, people who use their smartphones and blog posts as a modern day diarist might without considering income generation potential. For them, generating posts that attract large numbers of followers are reward enough. The simplicity of the technology of a smartphone camera with flip mode and easy to use post-production apps have provided the tools to help them connect with others.

### 2.3.3 *Moving Image Projects*

*Project 2-to-2*

A project students have found very useful for exploring the possibilities of a camera phone, as well as understanding the importance of good editing, is a variation of the *24 Frames 24 Hours* project devised by Max Schleser with over 200 co-creators (Schleser and Turnidge, 2013). *24 Frames 24 Hours* used the techniques of Cinema Vérité and became a global event linking countries around the world via a web-based distribution platform. Schleser set a specific framework for the project and curated an online gallery of work to showcase the growing network of mobile filmmakers. He suggests, 'By means of establishing a digital canvas online, participants can paint a picture of their local environment and communicate through the social media integration. As global collaboration is naturally related to the notion of time difference, *24 Frames 24 Hours* references this working parameter' (2013: 207). In *Project 2-to-2*, the emphasis is primarily on developing editing skills. Producing a work of exactly 2 minutes duration allows students to reflect on what is absolutely necessary to convey meaning in their narrative or montage. For the *24 Frames 24 Hours* project, Schleser emphasises the cultural aspects of filmmaking and suggests that by developing online communities innovations in creativity and "new forms of self-expression" may result (*ibid.*). Once you have completed the editing process of *Project 2-to-2* you might start to explore more fully the dimensions of collaborative work and what it means to be part of an online creative community for your practice.

---

**Project: 2-to-2**

Theme:

To explore your city, town or village from a personal perspective. You might consider how the camera phone can be used from a subjective point of view (POV), as an extension of yourself. Concentrate on the visuals rather than sound at this stage. A soundtrack can be added during post-production.

Method:
- You will choose a 2-hour time period out of 24 in which to film all your material.
- You need to show your start time and end time for filming. This doesn't have to be using a watch or clock. You can think creatively how you will demonstrate to your audience your time period.

---

(*Continued*)

*( Continued )*

- The audience for your work is the online global community.
- When you have completed your filming you will edit all the material to a running time of exactly 2 minutes.

There is free editing software which give you basic templates such as iMovie standard with Apple computers, Movie Maker for PCs or current Apps for mobile phones such as iMovie, FilmoraGo (iOS), PowerDirector (Android), that you can experiment with in the first instance.
For undergraduate students Final Cut Pro-X, Premier and Avid are standard in many institutions where it is expected you will progress to a more technically advanced and industry supported software.

- You can add text, inter-titles, sound and voice-overs for creative effect.
- You can add an appropriate royalty free soundtrack or make your own if you have musical skills.

What to do next:
Start your own online Project 2-to-2 community and share your work with each other.
Arrange a *Project 2-to-2* screening event.
Develop your project further and join the *24 Frames 24 Hours* community (Schleser, 2013).

## 2.4  Sound and the Archive

The focus of this book so far has been towards visual imagery whether moving or still. In this section, we will examine the equally important role of sound in creative mobile media and the voice memo and sound recording function of the smartphone. The sound recording capabilities of early mobile phones tended to be of very low quality. As a consequence much of the aesthetics around moving image work was developed by the production of silent films that had music and sound added in post-production. As the sound recording capabilities improved and with the appearance of external microphones that could be plugged into the mobile phones, so did the way sound was used in the works created. For students creating more advanced micro dramas and feature length films external microphones, plugged into the smartphone, for sound recording are usually the preferred option.

Sound recording capabilities of the smartphone continues to improve and it is interesting to note, particularly in the case of journalism, that it is in radio that smartphone usage has had the biggest impact. Journalists are using additional software applications that have boosted the function of the mobile phone by transforming sound into mobile IP-codec. They can record audio, edit and send it to the studio via a Wi-Fi connection. Marc Settle from *The College of Journalism* and *BBC Academy* comments:

> 'Almost daily, reporters are recording items which they then email direct to the BBC's playout systems; while others are using an application called LUCI Live to broadcast live, in quality, from wherever they are' (online, accessed April 2015).

One of the more interesting developments in sound has been in the recognition of the role of oral history and the archive for ordinary people. It is a revival of projects that have their roots in The Mass Observation Archive (http://www.massobs.org.uk/, accessed August 2015). Between 1937 and 1950s the Mass Observation Social Research Organisation collected large amounts of data from ordinary people and gave a snapshot of the country at particular moments in time. The information was collated largely from personal writings, diaries and surveys but as a longitudinal study it was able to track changes to 'everyday life in Britain'. For the first time there existed large amounts of information about the lives of ordinary people.

The National Sound Archive, formerly The British Institute of Recorded Sound, was founded in 1955. It holds over 550,000 hours of sound ranging from early recordings on wax cylinders to digital formats and covers a range of genre from drama, wildlife, music, literature to oral history (Robinson, 2005). Part of the oral history section contains recordings from the 1950s covering an 11-year period of data collection as part of the Survey of English Dialects. Vernacular forms of speech and traditional dialects have been preserved from over 304 recordings (*ibid.*). It is also through listening to these preserved voice recordings in the archive that, historians have been able to trace the migration and decline of regional accents and dialects around the country. The National Sound Archive is held at the British Library, which has been behind a number of recent initiatives that focus on sound recordings. Unlike previous sound archives which relied on experts and trained volunteers to make the recordings, the digital environment, including smartphones and the Internet, have made it possible for more of the general public to be directly involved.

In 2014, BBC Radio 4 first started broadcasting *The Listening Project*. This was made in conjunction with the BBC's national and local radio stations and the British Library with a remit of "capturing the nation in conversation". People across the UK were asked to share 'an intimate conversation with a close friend or relative' and be part of the creation of a national archive that was to be preserved for future generations (http://www.bbc.co.uk/radio4/features/the-listening-project, accessed April 2015).

The *Listening Project* was inspired by an American version started in 2003. *StoryCorps* became one of the largest oral history projects with over 80,000 participants taken from every American state with 50,000 interviews undertaken. The archive was free to share and preserved at the Library of American Congress and the American Folklife Centre (http://storycorps.org/, accessed April 2015).

A second sound project introduced in the UK in 2015 was *The Sound of Our Shores,* a joint scheme between the British Library and The National Trust. It was launched on the BBC website as a community led project to create an audio archive of the UK seaside. The public were asked to contribute by recording their favourite seaside sounds. The sound clips will be uploaded to a digital map of the country via the *Audioboom* website and form a permanent record. As such, a constantly changing coastline is being preserved as a "snapshot for future generations" (http://www.bl.uk/sounds-of-our-shores, accessed April 2015).

These initiatives have not only created a wider interest in the sound archive but are focused on the collection of national resource for future generations to access digitally. Both *The Listening Project* and *Sound of Our Shores* demonstrate the versatility of the mobile phone in your pocket and what can be achieved through sound recordings whether they be of the natural landscape or conversations.

Another sound project that made for interesting listening and contributed to a growing body of archival material of the British Isles, was *Evening Echoes* (1993–1995). Artist John Carson and musician, Conor Kelly travelled the length of Britain and Ireland recording the cries of newsvendors as they advertised the local evening papers through very distinct and localised calls. The soundscape was exhibited as part of an installation alongside life-size photographs of the vendors. A CD was produced based on the street sounds, intonations and regional accents of the newspaper vendors from 23 of the cities visited. Once a familiar sound landscape to those dashing home from work in the evening the loss of many local newspapers to online journalism platforms and changes to British accents and dialects means these cries are no longer heard and the recordings become part of memory and history (http://artscool.cfa.cmu.edu/~carson/evening-echoes.html, accessed March 2016).

The sound projects illustrated in this section engage with differing aesthetic styles which are specifically driven by the context of their reception, within the gallery space, on radio and online. *The Listening Project* functions across both radio as a purely sound medium and on a website that includes both sound clips and visual and written material. We might consider how digital technology can be used in creating such work and the way it informs our choices for making a personal sound archive. Social media is largely about a visual culture and operates like a form of extended family album, albeit with a global reach. We still use photographs as an aide memoir and can search images of family members from generations long past. It is unlikely that, for most people, there is such an archive of personal voice recordings. Very few families consciously preserve voice recording of grandparents, parents and children in the same way they do with visual records even though the technology in the form of tape, cassette and now digitally was readily available. Perhaps, it is with the growing awareness of the importance of sound to trigger memories and the increase of conditions such as dementia and Alzheimer's that interest in the sound archive is growing. The following exercises are designed to get you thinking how you might use sound to develop your own archive, soundscapes and sound-stories.

---

### Sound Projects

1. Consider how you might use the sound functions of your mobile phone to start your own version of a listening-style project. What conversations would you like to share with others?
2. Create your own soundscape from places you have visited including: *Sounds of the City* or *Sounds of the Seaside.*
3. 2-to-2 Sound Project: Make a sound 'picture' or journey in your village, town or city following some of the rules from the previous 2-to-2 project. This time, create your visuals in your head through listening to your soundscape. To help you develop the essentials of editing it is important you edit your 2 hours of material to exactly 2 minutes running time.
4. Interview family, friends, grandparents and/or people in your wider community. Create your own archive.
5. Echoes — Is there something in your local area that is about to disappear and can be captured in sound?

---

### Some examples of documentary films

The danger of creating lists is that many things get left out. Below is a selected but by no means definitive list of documentary films to view. They demonstrate a range of styles and techniques from some of the first documentaries to contemporary practice.

(*Continued*)

*(Continued)*

Robert Flaherty *Nanook of the North* (1922, USA)
Dziga Vertov *Man With a Movie Camera* (1929, USSR)
Errol Morris *The Thin Blue Line* (1988, USA)
Michael Moore *Fahrenheit* 9/11(2004, USA)
Werner Herzog *Grizzly Man* (2005, USA)
Jennifer Baichwal *Manufactured Landscapes* (2006, UK)
Molly Dineen *The Lie of the Land* (2007, UK)
Bartosz Konopka *Rabbit a la Berlin* (2009, Poland)
Kim Longinotto *Pink Saris* (2010, UK)
Stevan Riley *Fire in Babylon* (2010, USA)
Adam Tyler *Kingsmeadow* (2011, UK)
Kevin Macdonald *et al. Life in a Day* (2011, USA)
Dror Moreh *The Gatekeepers* (2012, Israel)
Gabriela Cowperthwaite *Blackfish* (2013, Iceland)
Jehane Noujaim *Al Midan* (*The Square)* (2013, Egypt)
Asif Kapadia *Amy* (2015, UK)
Richie Mehta, Mandira Chauhan, *et al. India in a Day* (2016, UK)

# Chapter 3

# Promotion and Exhibition

In this chapter, we examine the range of platforms for dissemination of media and consider how to promote and exhibit work. The annual iPhone and Mobile Phone Film festivals and the Mobile Innovations Network Australasia (MINA) are just some of the physical and virtual spaces for screenings and networking on a global scale. Social media sites such as YouTube, Vimeo, Tumblr and Instagram create possibilities through their distribution channels for sharing work with others across the globe. At a more local level, exhibiting in a *pop-up* gallery or producing work for a charity may be more appropriate for some practitioners. So, how do you begin to promote yourself, your ideas and develop your 'brand'? This chapter examines what creating projects for an audience means and the variety of ways of getting visual and audio material into the public domain while keeping work legal.

Let us start by considering the following statement: 'Exhibitions are strategically located at the nexus where artists, their work, the art institutions, and many different publics intersect' (Marincola, 2006: 9). Marincola focuses on the relationship between creating a media project and considering how work is disseminated and to whom. Not everyone begins a film or a photography project with a pre-existing client, brief or outlet for the work. Social media sites have made it easy to get a film clip into the public domain without necessarily thinking about how it may be received. The intention here is to look at some of the options available to you, including social media, with a more critical eye so that you will be able to make informed choices about the various platforms for distribution as well as having an understanding of what is appropriate for your specific project.

You were asked to think about some key questions before embarking on any production project that included considering what you want to say, how you want to say it and why? The *what* engaged you with both practice as research and research through practice. The *how* is asking you to consider methodology and the techniques you will draw on in creating your media project including the research of other practitioners. The *why* asks you to consider the purpose of the work and whom it is created for, that is, your audience. By starting to formulate answers you will have already selected a particular medium to work in. You will have considered some of the conventions and codes of construction applicable to that medium and already be familiar with some of the platforms available to share work with others. For example, sound projects may be linked to radio production and a broadcast audience or be part of an online website or downloadable podcast. Photography can be exhibited online or you can make physical prints that can be viewed by an audience in a gallery space. Posters are part of an integrated media encompassing graphic design and photography. They can be digital or printed and can be displayed in numerous locations from the art gallery to a bus shelter. These different forms engage different kinds of audiences and in turn shape how those audiences experience your work.

Marincola suggests that by looking at questions of and around practice the emphasis is placed on 'how concepts surrounding curating are filtered through the lessons derived from repeated performance, from thinking *and* doing, or perhaps more accurately, thinking based on doing' (2006: 10). By considering the structures, methods and conditions of exhibition practice she draws attention to fundamental questions about the nature and role of the curated work. In a sense the exhibition is about the dissemination of information and/or knowledge. Even work that may be put together in an arbitrary way involves a selection process. Greenberg comments: 'Exhibitions and anthologies are, by definition, selective and exclusive due to the biases of the organisers and the actual or perceived constraints of space, finance and availability of works' (Greenberg *et al.*, 1995/2009: 2). These are issues that need foregrounding if one is to take a more critical approach in considering how work is exposed to the wider public and this applies as much to online distribution as it does to the more conventional exhibition spaces. So, an awareness and understanding of the context of exhibition and distribution are fundamental to making informed choices about your work. There can be positive aspects to the challenges raised once you start to critically engage with such issues.

For example, whilst budgetary limitations may have a considerable effect on the final outcome of the exhibition design and marketing these can also lead to creative and innovative ways of overcoming monetary constraints. There might not be the budget for poster printing costs but time may allow for a *spreadable* online campaign. The emergence of self-funding exhibitions, pop-up galleries and group shows have made it possible to set up exhibitions more cheaply than the conventional route of hiring out established gallery spaces or persuading major sponsors to cover the costs. There are also online fundraising platforms with a rationale to help creative people launch their projects.

*Kickstarter* was launched in 2009 and is a popular online platform and resource that hosts a range of creative projects from films, music and art, to games, design and technology. Projects are independently created setting funding goals and deadlines. As the site explains: "If people like a project, they can pledge money to make it happen" (https://www.kickstarter.com/, accessed July 2015). One such example is a final year student project undertaken as part of undergraduate studies in media production. A mobile documentary film was created to introduce the launch of a music band and used the Kickstarter platform as a way of both crowd-funding the release of their first album and also to provide visibility for the documentary that was part of a wider promotional campaign package. It brought an additional, meaningful dimension to a project created as part of a final year university degree and crossed-over to a more public facing setting. This is something all final year undergraduate students might consider when planning practical work. If you can incorporate aspects of your work that include an outward looking approach it may be useful in getting you and your ideas noticed; enable you to develop a network of contacts with experience who may be able to offer support in the future. The creative and media industries are difficult to break in to but everybody has to start from somewhere and showing work to others is the first step.

Producing work with a mobile phone doesn't have to be any different from that created using professional level technology when it comes to exhibition. Moreover, some of the considerations discussed in this chapter will apply to any kind of art practice. However, it is worth considering the difference between work created on a mobile phone and work created for a mobile phone. The latter is for viewing on a very small screen and though by definition is mobile, it is also for the solitary viewer. Furthermore, if sound is part of the work, it is normally heard through headphones or ear buds. Finally, it may involve the audience being mobile and moving within or between particular locations. For example, if you are using ideas based on the methodology of the *flâneur* (see Chapter 2) you might want to recreate the experience of observation and contemplation within the city landscape. For those producing work in sound only, your aims might be to create an immersive experience. Michael Bull examines the notion of the "sound bubble" created through the use of iPods and MP3 players when he suggest: 'Users can traverse the noisy urban scene empowered precisely through filtering out the sounds of the city, to be replaced by their own private sound bubble provided by their own choice of music. Users frequently embark upon a range of strategies of sonic transcendence. Central to this is the construction of urban space as a "seamless space"' (2013: 13). The smartphone merges the functions of the MP3 player, the phone, camera and Internet and allows for multiple sensory experiences whilst filtering the sounds of the city. It is possible to think creatively about how the audience might receive the work and their relationship to technology. A podcast allows listeners to download work onto their personal handsets and play it back at a time and place that suits them. A different kind of sound project may direct them to a specific place and work in relation to the surroundings as a form of augmented reality (AR).

## 3.1 Organising an Exhibition

### (a) *The Audience*

What is your rationale for creating a piece of work and why do you want to share it with others may seem an odd questions to pose. Many creative people would argue it is something they just have to do whilst others may see it as a way of becoming famous or making money. There are numerous reasons for wanting to do something but thinking through your rationale and ambitions are important considerations. At the start of this book I appealed to students to extend themselves and consider how they might use work to benefit others. Entertainment and pleasure may be just as valuable considerations as more 'worthy' causes. When I speak about audience consideration it is not to constrain or prevent you from doing one kind of project over another. Neither am I suggesting that audiences always have to like what you have created. Some of the most controversial documentary films were not made to be liked but to compel people to see things from another point of view or perspective. Good communication is essential. Your audience needs to understand at least some of your aims and intentions or be moved by the work in some way for them to have a meaningful experience.

You do need to consider what your responsibilities are when making work that goes into the public domain and that you work ethically and within the law (see Chapter 9). If you are making a film with explicit sexual content or violent scenes, both the age group of your audience and the place of screening not only matters but is determined by a set of rules and guidelines in the UK developed by the British Board of Film Classification (BBFC available at http://www. bbfc.co.uk/what-classification/guidelines, accessed April 2016). The BBFC have on their website information about digital and mobile content that is worth looking at when planning your projects in the UK. Other countries will also have laws and guidelines about what you can do or show in a public space and which you need to familiarise yourself with if exhibiting outside the UK. In a gallery setting work unsuitable for those under 18 years of age is usually separated from the rest of the exhibition space and viewers alerted to the content through signage or stewardship.

The purpose of the first two chapters was to help you explore some of the historical foundations and conventions that underpin media work and how shared meaning is both created and circulated. In considering the audience you will further explore how some usage of language becomes privileged over other forms of expression. For communication to be effective it should enhance the experience of the viewer. Deciding to exhibit your work draws attention to both the work and to you so knowing how to present work and engage an audience is one of the first requirements in becoming a media practitioner.

## Gallery: Visit 1

It is at this point that the students in my class are sent on their first programmed visit to explore a gallery with a focused set of research questions. The purpose is for them to start thinking about what they have learned so far and to consider how more established galleries do things. By engaging critically with what you experience when you visit a gallery you can gain insight and apply the learning to what you would like to do and perhaps more importantly, what not to do.

It is important to visit some of the large exhibition spaces where, because of sponsorship, budgets are less constrained. Such galleries can attract a more diverse range of exhibitions covering all aspects of media work. They also utilise space differently and have a different approach to the written material and labels they provide alongside their displays. Consider the differences and purpose between public and private funded galleries and local, city or national galleries. A national gallery belongs to the nation but as the introduction to this chapter suggests, what is exhibited has been chosen through a process of selection. One might therefore consider both the cultural and symbolic functions such a place serves in relation to its historical origins and its continuing role as a centre for representing the nation (Boswell and Evans, 1999). Bennett suggests: 'Museums, galleries, and, more intermittently, exhibitions played a pivotal role in the formation of the modern state and are fundamental to its conception as, among other things, a set of educative and civilizing agencies' (1999: 338). This raises further questions about the historical function of national institutions of culture in constructing and maintaining nationhood. Bound up in such functions, is having a repository of cultural and symbolic artefacts that can showcase the nation to the rest of the world in a particularly selective way. Educative and *civilising* functions have long been regarded as contentious and became part of a wider discourse on power, class, race and inequalities. We might ask if such functions are still relevant in a postmodern, globalised, interconnected world and whether the idea of nationhood is still fundamental to the way both individuals and countries see themselves and their place in the world. This might form the starting point for a classroom discussion. The United Kingdom's national galleries and museums are thriving and part of any visit might be an evaluation of the exhibits in relation to the cultural and symbolic meanings that underlie them.

Figure 3.1 provides guidelines and key areas to consider on your visit that focus on both the exhibits and the way the gallery space has

---

**Gallery Visit 1**

Below are some headings to help you focus your visit. You will be engaging with the exhibitions, artworks and displays but it is important to consider some of the organisational details that make the experience enjoyable or not and how it will inform the choices you make for exhibiting your own work.

- **Entrance and Exits**
Are they clearly visible? Is there circulation space? Are they level/is there disabled access?
- **Seating and Rest Areas**
Where are they located — inside or outside the exhibition space or both? What kind of seating? Bean-bags, hard chairs and tables? Is there a café or refreshment area? Toilet facilities?
- **Information Signage**
How are these displayed and where? What information do they carry? Are there other visuals and graphics?
- **Gallery flow and routes through exhibitions**
Consider how you navigate through the exhibition space. Can you go at your own speed? Can you spend as much time as you want in the space? Are you directed in a particular way? Could you miss parts of the exhibition? Collect any printed guides or maps that may be provided.
- **Displays**
Size of labels and panels? Are they close to the object or is their placing confusing? What typefaces are used? How is upper and lower case used? How is colour used? Is there enough contrast with the background for clarity? Are there any overarching narrative themes explicit or implicit in the way the exhibits are organised?
- **Audio/Visual**
What special technical equipment is used? Do films run on a loop or can the viewer activate them? How are screens/monitors mounted on the wall? How are images projected and on to what surface?
- **Lighting**
How are the exhibits lit? Is there glare, reflections or shadows? Does the lighting change as you move around the space? How are entrances/exits and stairs lit? Is text and signage lit? Is there daylight in the gallery space? If so how is this used or controlled?

Please feel free to add any additional notes you think will be useful.

---

Fig. 3.1   Gallery Visit 1.

been used. Even if you are regular gallery visitors it is worth considering this from a new perspective and be open to new experiences.

### *Gallery/Location: Visit 2*

The first gallery visit should have armed you with a breadth of knowledge about how exhibitions are put together, how space is used. If you have managed to visit a large established gallery, you will have noted the additions that come as standard and form an important part of the experience, such as the cafeteria and gift shop, which are also a source of income generation. The second visit should be centred on the specifics of exhibiting your own work. To this extent find a venue that you think would make a good space. You could start with a smaller gallery but equally it is well worth exploring spaces such as your local café or the foyer of a hospital or library or even a disused high street shop. Such spaces have great potential for displaying work and are increasingly being used for pop-up exhibitions. Apply some of the same criteria you used on your first visit when considering these kinds of spaces but bear in mind you will also need to engage with the owners of the properties if your plans are to come to fruition. If a venue has already been used for displaying the work of artists then this is an indication that the owners are supportive of such ventures and are used to being approached.

Pop-up spaces are regularly advertised on the Internet and below are listed some examples from the UK that provide information on spaces that are available at any given time. These spaces are transitory by nature and can disappear at a moment's notice. The sites also provide information about the charges involved which you will need to factor into your budget.

Artquest:
    http://www.artquest.org.uk/articles/view/empty-shops (accessed April 2016).

London Pop Ups:
    http://www.londonpopups.com/p/advice-resources.html (accessed April 2016).

Promotional Spaces:
  https://promotionalspace.com/ (accessed April 2016).

Pop Up Gallery:
  http://pop-up-gallery.co.uk (accessed March, 2017).

Pop Up Brighton:
  http://www.popupbrighton.com/ (accessed April 2016).

## (b) *Situating Work within an Exhibition Space*

Student work is usually exhibited as part of the final year or end of year show that both encompasses the safety net of a university group show and allows selected audiences to engage with the work and offer valuable feedback. Depending on the institution you are part of these can be casual, informal affairs or highly professional exhibitions and screenings that use a variety of venues on and off campus. There is often a departmental budget and a marketing and publicity team made up of experienced staff to support such events. You will also have the expertise of technical staff to support the display and presentation side of things. You can learn much from playing an active role in such events and many students use these opportunities to form their first networks of like-minded people. It is no surprise that some of the most successful contemporary UK artists moved on from their university or art school shows to curate their own group shows. In 1993, British artists Tracey Emin and Sarah Lucas set up a temporary gallery and social space, *The Shop* in Brick Lane, London to display, promote and sell their work (Abbot, 2013). From the publicity generated, which attracted sponsors, gallery owners and other artists, the two went on to successful careers as part of the Young British Artists (YBA) movement along with the likes of Damien Hirst, Gary Hume *et al.* and contributed to the shaping of 25 years of contemporary art in the UK.

The second field visit should have enabled you to identify the kind of space most suitable for exhibiting your project and what is available in your local area. Organising the exhibition yourself gives you an insight into what is involved. More importantly, it gives you the control to show the work in the way that you want. If, at a later date, you are successful enough to be invited to exhibit at a more established gallery where teams are employed to do most of the work for you, you will have gained a better understanding of what you want from this experience (Duffin, 1991).

When we think about exhibition space we need to consider every aspect of how it could be utilised. For example, it can be envisioned as 'an environment that communicates' and that requires a balance between space, objects, information and technology (Lorenc *et al.*, 2010). Exhibition design, therefore, is about utilising the environment whether it be physical or virtual, so it is important to consider how it can be used to create a stimulating experience. It is also multi-disciplinary in that it involves considering the architecture of a building, the lighting, interior design, graphics, digital media and audio. Lorenc *et al.* use the analogy of composer, orchestra, choreographer and conductor working as a team to produce a meaningful experience (*ibid.*).

Exhibition spaces can be small, intimate venues or cavernous warehouses. When considering space it is important to think about how you want people to move through it, known as the *flow*. Can they choose which exhibits to look at as and when they want or have you, as the curator, directed how people circulate through the space? Different routes through an exhibition space can create different experiences for the audience. So, where are the entrances and exits of the exhibition space situated? These may determine how you direct the flow of people. If you have one door that serves as both entrance and exit how is it managed? What safety considerations do you need to be aware of? This could include displaying clear signage for emergency evacuation in case of, say, a fire. Depending on the kind of exhibits you intend to have you may need to consider seating arrangements in your space or screens that divide the space into specific areas. In managing the space you are creating an environment for your audience and providing them with all the conditions to create a memorable experience.

These ideas can be taken further by thinking about the concept of *universal* design as noted by Lorenc. 'Gone are the dark days when public venues could only be enjoyed by people with certain abilities.

Title of the project:

Name and address of the producer/production company/photographer:

I (name) have understood the information given to me about the project (title) as provided in the information sheet (dated) and my role in the work.

I (name) hereby consent to the filming/recording and participation in the work: (title). We grant (name above) the right to use my likeness, voice, performance, biographical details in the making, exhibiting and distribution of the project titled: (name) including all rights for any media formats including online distribution in perpetuity.*

I consent to (name) having sole rights to use my name, likeness, biography, photographs, recordings, interviews* to publicise the work in all media formats and throughout the world.

I confirm I am over 18 years of age.

Signature:

Date:
Address:

Telephone:

Fig. 3.2   An example of a simple participant consent form for a film project.

* These are general headings and can be deleted or exclusion clauses added during the negotiation process. Please note this information is given as a general guidance to the kind of release forms you will find online. Please make sure you research this area fully and provide an appropriate release form for performance/music/location as needed.

The concept of designing for "the handicapped" or "disabled" should also be regulated to the dust-heap of history' (2010: 18). What Lorenc is implying by this is that universal design is, from the outset, inclusive. It allows experience and opportunity for all and includes a recognition that signage may need to be in braille and/or "layered" for different levels of reading ability; areas should be accessible by wheelchair, for example. Public institutions have responsibilities to make exhibition spaces accessible to all. If you are designing a space by yourself and have a limited budget it may not be possible to cover every aspect needed to ensure accessibility for all but approaching the task from values intrinsic to universal design should be an ideal to strive for.

Consideration of space also extends to how work is exhibited in the gallery. For example, photographs can be placed in frames within a card window mount. This forms a specific, pre-determined space around the image. Unsurprisingly, common convention means the mounts tend to be in neutral shades; cream, white, grey or occasionally black. Mount colours are chosen so they enhance the image rather than detracting from it. How photographs are displayed communicate subtle messages to your audience. They don't have to be in frames. Contemporary images can be mounted on a variety of materials including fabric, metal, wood, glass, or on light boxes. Mobile media exhibitions may utilise mobile phones, tablets, flat screen monitors or projections in the gallery space to offer a more interactive experience.

## (c) *Lighting*

Exhibition lighting is a specialist area of design that may need to deal with challenging situations. This applies particularly to the larger galleries where several curators are mounting simultaneous exhibitions. In these cases, striking a balance between those designers who like to use daylight and natural lighting and those who prefer artificial lighting, either to preserve delicate objects or highlight exhibits, is no simple task (Lorenc, 2010). If you are renting space in an established gallery the lighting design may already be in place with strict limits placed on what you can change. The more transient, pop-up venues also present challenges, either in terms of what the space allows or what the budget will allow.

Lighting, like colour (as discussed in Chapter 2) is mood altering so you would be wise to start by considering from the outset what kind of experience you want for your audience. How do you want them to feel? Daylight, particularly sunlight, is said to have a positive effect on mood but sunlight can also damage artefacts, bleaching photographs, paint and fabrics. If screening a film is also part of the exhibition you will need to consider how you might manipulate the light levels so that the audience can see the images as you would want them to. Reflective surfaces can also present a challenge, for example, when photographs are hung behind glass. If you can position spot-lights in a way that diffuses the light this can help. Spending time to assess the lighting conditions of your venue and experimenting to find the best possible combination of light is key to showing work at its best and so creating a memorable experience for your audience.

### (d)  *Risk Assessment, Health and Safety*

If you hire out space at an established venue insurance cover is usually included as part of the fee and the establishment may well have carried out a basic risk assessment and have clear guidelines as to what has to be done in the event of an emergency. There will be members of staff on hand who are knowledgeable about health and safety issues and you will need to comply with their regulations. If you are using a pop-up venue you may have to organise these things by yourself. Undertaking a risk assessment forces you to consider the things that might be of high, medium or low risk to anyone coming into the venue and for-mulate a policy for dealing with the hazards. For example, will there be anything already in the venue or be present as a consequence of the exhibition that may constitute a fire hazard? Are there flammable materials and if so, what might you do to lessen the risk of them catch-ing fire? You may be required to spray any such materials with fire retardant before being able to use them. Exits will need to be clearly marked and a strategy put in place for the safe evacuation of the build-ing. It should be possible to call the emergency services at all times should the need arise. Being mindful of precisely what constitutes a hazard and how to reduce risk needs to be part of good working prac-tice particularly where other people, the public, are involved.

### (e)  *The Group Show*

The group show is often one of the first ways those new to exhibiting will be able to get their work seen by the public. Costs can be shared and there can be consistency across promotional materials and publicity. A good show will aim to link the work on display in some way, thematically or otherwise. Ralph Rugoff notes, 'For better or worse, our experience of art is not exempt from our susceptibility to the power of packaging. And a themed exhibition is ultimately a type of packaging' (2006: 45). In this statement he isn't just making reference to commercial packaging with its aim of grabbing our attention in any way possible but in having some influence in how we ultimately make sense of the work on view. This may be to provoke or question what we experience. Individual work may vary in appeal but the context should invite the audience to an enriching experience through the use of considered juxtapositions and negotiation of space. In this respect, consideration needs to be given to how the audience will experience the event as a whole along with how they negotiate individual artworks or screenings and whatever else may be provided for their comfort in the way of extra facilities.

Even in a pop-up culture rigorous organisation and planning are required before an exhibition can open to the public. At the very least the location or space will need access to toilets. You might also consider providing refreshments for the opening night, which immediately begs the question, is the venue permitted to serve alcohol? Some venues may require a temporary drinks licence and applications for this take time and planning.

## (f) *Moving Image*

The traditional exhibition space for film and moving image work has been the cinema or television. From the 1960s onwards, however, it started to have significant presence within the art scene (Nash, 2006). Now it is common for gallery spaces to offer the opportunity for work  to be projected, in multi-screen formats if required, and have a co-presence within specific spatial arrangements. Gallery screenings are very different from traditional cinema screenings in that are consecutive or if screened simultaneously, are confined to very discrete spaces. They utilise both small and large screens and these may also have to compete with each other for the audiences' attention. Sound can either permeate the entire gallery space or through the use of headphones allow for intimacy and personal engagement. How you want your audience to experience the work is of prime consideration when thinking about location and space whatever the medium.

As with all the things discussed so far in this section, the budget is an integral factor. What technology pre-exists in the space you have use of or what might you have to hire? If you are planning pop-up screenings then your budget will dictate the technology you can use in the space you have access to. Mobile technology is increasingly becoming part of the user experience in galleries and interactive work, so tablets, mobile phones or computer screens, will need secure anchorage. This not only allows your audience to use the technology safely but also minimises theft. The devices are also able to provide additional and detailed information on the artist or works that can't be accommodated in the primary exhibition space. Moreover, some established artists are embracing such technology to experiment with and extend their practice with what for them is a new medium. For example, David Hockney's 2012 exhibition *A Bigger Picture* at The Royal Academy, London included, alongside more traditional paintings and drawings, a room dedicated to work both created and displayed on iPads (Barringer *et al.*, 2012).

## 3.2 Film Festivals and More

Many filmmakers producing short films rely on the festival circuit to get their work shown. The large international festivals, Cannes in France, Berlin in Germany, Venice in Italy and Sundance in the USA are likely to be familiar to many interested in mainstream cinema. They generate extensive media coverage and attract leading filmmakers, film stars and other celebrities. A more productive avenue for the self-funded filmmaker is to consider participating in one of numerous smaller, niche or more localised festivals. There are three main functions that a festival can offer to a relative newcomer. Firstly, they provide an opportunity to have your work screened to an audience that tends to be both passionate and knowledgeable about film.

Furthermore, the question and answer sessions, which are part and parcel of film festival screenings, hone your skills in articulating your ideas. Being able to answer sometimes challenging questions contributes to confidence building and making interaction with the public a less daunting experience. Secondly, festivals act as a market place. If you are lucky your films may be picked up for wider distribution. Thirdly, it is an opportunity to meet like-minded, creative people at different stages of their career and so widen your networking circles. So, to summarise, festivals are a learning opportunity where you can gather information, gain experience and be exposed to new ideas alongside screening your film.

Established film festivals usually include a short films section and some have extended the film category to include mobile media. All festivals these days have an online presence and some of the key mobile festivals are lively virtual spaces. The following are just some of the mobile festivals taking place around the world, however, it is important to do your own research and find the ones you feel are most appropriate to showcasing your talents.

The *Syrian Mobile Film Festival* deserves a special mention. Syria, a country in the midst of a civil war, held its first Mobile Film Festival in 2014. It was started by a group of filmmakers and activists and perhaps embodies best what creative practice and collaboration can achieve. The importance of mobile media as a tool for transformation and change is highlighted on the festival website where filmmakers often risked their lives to produce the work and stage the event. The mobile phone is described as 'the main tool for peaceful struggle and free expression' (Syriamobile Films, 2014). The first festival took place over a 3-day period in October 2014 across five Syrian towns. The publicity generated from the first event along with a website extended its reach from the local to the international. Alongside the screenings and political activism of what has become an annual event, work continues throughout the year with the educational platform, *Pixel*, offering training programmes and workshops for aspiring filmmakers. More information and some of the award winning films from 2014 to the present are available at http://syriamobilefilms.com/en/ (accessed June 2016).

The *iPhone Film Festival* (IFF) was started in 2010 and attracts mobile filmmakers from around the globe. There is a list of rules for those submitting films including one that stipulates the use of a mobile devise, iPhone, android or GoPro camera, for at least 70% of the material in any film. IFF places restrictions on the submission of films to other festivals and has strict rules regarding the length of films submitted. Prize-winners are selected from different categories of films by a jury. Further information is available at http://www.iphoneff.com/ (accessed June 2016).

The *Pocket Cinema Film Festival* is described as 'Asia's biggest smartphone festival' and originates in Pakistan (http://pocketcinemaff.com/, accessed June 2016).

The festival takes place over 2 days and includes both competitive screenings and a conference along with a website that showcases many of the films.

The *Mobile Film Festival*, based in France, is in partnership with the United Nations for its 11th edition. There are juries awarding prizes and the usual sponsors of such events can be found linking many of the mobile websites. Further information is available at http://www.mobilefilmfestival.com/ (accessed May 2016) and provides a viewing platform for the work to be screened.

MINA co-founded by Max Schleser and Laurent Antonczak is now in its sixth year. It comprises an annual symposium and screenings held at various institutions in Australasia and has an online presence at http://mina.pro/. It is described as 'an international network to promote cultural and research activities that examine and expand the emerging possibilities of mobile media in New Zealand and internationally. MINA aims to explore the possibilities of interaction between people, content and the emerging mobile industry' (Schleser, 2013: 105). As an international network it offers far more than a film festival and its collaborative innovations have been far reaching. It was one of the first spaces for academics, practitioners and industry to come together both physically at the symposiums and equally as importantly, in its virtual spaces. The short film *May Days* discussed in Chapter 6 had one of its first public screenings at MINA followed by another of my mobile projects *Office no. 47* in 2012. In 2013, the peer-reviewed *Ubiquity: Journal of Pervasive Media* featured a special Mobile Media edition with Schleser discussing MINA and other topics pertinent to mobile culture and communication.

Originally created by Airing Korea TV as a cultural event for Korean–Canadians (AKSSF), the Toronto Smartphone Film Festival (TSFF) had a name change in 2014 to reflect a wider, inclusive remit. The international festival runs an annual 2-day event open to all Canadian and international mobile filmmakers. Film submissions are limited to 10 minutes or less duration and terms and conditions apply including allowing 'unrestricted use of the film for any purpose, without compensation to the provider of the film.' Information is available at http://www.smartphonefilm.ca/ (accessed June 2016).

*Raindance* is one of the largest independent film festivals in the UK. It was started in 1992 by Elliot Grove to offer practical, filmmaking classes and the following year the first film festival took place. The festival is an annual event held in London over a 12-day period and screens films from around the world. It is officially recognised by the British Academy of Film and Television Arts (BAFTA) and in the United States by the Academy of Motion Picture Arts and Science. The festival accepts submissions in various categories including feature films, short films and music videos and charges a submission fee. Further information is available at http://raindancefestival.org/ (accessed June 2016). There is also a *Raindance Web Festival*, https://filmfreeway.com/festival/RaindanceWebFest (accessed June 2016) that offers opportunities for films created using mobile technology and streamed online.

The rationale for starting many an independent film festival has been to provide an alternative space to the established festival circuit. The top international film festivals can be highly competitive environments and difficult for new and less established filmmakers to get work screened. The smaller, more localised festivals have proved invaluable in offering independent producers or aspiring filmmakers a chance to showcase work that otherwise might not be seen. In such spaces the filmmaker can discover what has communicated well to an audience and what they may need to work on for future development (Elsey and Kelly, 2005). The themes and categories films can be submitted to vary and may include a wider choice than more mainstream festivals.

As these alternate festivals grow in popularity they start to attract greater funding from sponsorships deals that can reduce their uniqueness. Sponsorship may well provide much of the revenue to run the physical festivals but sponsors also make demands that could be onerous; at the very least they seek visibility through logos and advertising. Some have moved from open and free submissions to a selection process with panels or juries deciding what is screened. Others reproduce the award system of established festivals with winners for a specific category of film and prizes. This is not always the case so good research is advisable. Always check the criteria and submission details fully. Is a fee charged for applying? How is the film to be submitted? Consider what rules apply for the acceptance of your film in the festival, including any clauses waiving your rights on how and for how long the film is used. However, it is normal to give consent for the use of the film or parts of it for publicity purposes, including on the festival's website.

## (a) *The Film Club*

If festival submissions seem a little daunting, joining or setting up your own local film club may be an easier starting point. A local cinema might be persuaded to screen film club projects outside of normal opening times; digital technology has expanded the environments where work can be screened. Whilst there is nothing to replace watching the physical reaction an audience has to viewing a film, moving image or other media work, the Internet makes possible a unique and different kind of experience. If obtaining a venue for your film club is not an option it is perfectly possible to form one that is online. Hosting sites such as YouTube and Vimeo make it possible to have your work seen and critiqued by others. Participation is the rationale behind many online communities which tend to be inclusive and if they have criteria for selecting work, it is often thematic. This applies not just to mobile films but also to photography and art in general. If you find your work is not selected for a film festival or gallery show you can set up your own show online; start a digital film or photography club and invite others to join you.

As your work develops you can think about creating your own portfolio web pages. You might include a comments section so people can leave feedback. There are sites which, for an annual fee, provide templates that enable you to choose a format that suits your needs and that also include everything necessary to get a working website published in just a couple of hours. If you wish, you can register a unique domain name although this will also involve paying fee. An example of this type of service is *Clikpic* with information available at http://www.clikpic.com/ (accessed June 2016). If you are familiar with *WordPress* then *Photocrati* may be worth a look. Further information is available at http://www.photocrati.com/ (accessed June 2016). *Format* is another site with information available at: https://format.com/ (accessed June 2016). Please remember there are a great many hosting companies and it is worth spending time to find one that suits your needs and offers you a good deal.

## 3.3 Marketing and Promotion

Whether you choose to exhibit your work in a gallery space or share it with others online, at some point in the process you will need to think about promoting the work and yourself as the artist or practitioner. All the hard work of creating a project will be futile if no one gets to see it. Strategies that get your work known and make you more visible are worth engaging with and in this section we turn to some of the ideas from advertising and PR that can be helpful. The world of advertising and marketing is changing rapidly and web-based promotional strategies sit alongside the more traditional ones.

### (a) *Branding Your Product and Yourself*

'Being able to talk about why you make the work, what it is about, what interests you about the ideas you use, why you use the materials you do and how the work is arranged in a gallery setting, helps you to be more in control of how people read and understand the work' (http://www.aarts.net.au/resources/, accessed June 2015). This is good advice and there are numerous online resources such as this one at *Accessible Arts* that provide information about how you can promote yourself and your work for an 'inclusive' audience.

You can think about this in advertising terms and creating a *brand*. This means branding your production and yourself. A brand is a way of distinguishing yourself from others and branding involves creating something recognisable and visible (Healy, 2008). Together with branding, promotion is part of generating visibility about yourself as the creator of the work. The idea of personal branding has its critics and *re-packaging* or *re-designing* the self for success can be seen as part of a strategy derived from PR, consumer and celebrity culture and social media that extend branding into both the personal and the work place. Changes to society that have altered employment from stable, permanent and long term to a flexible, mobile and casual form of labour require new thinking where personal attributes, skills and adaptability of the individual are viewed as assets. These economic and cultural shifts are seen as part of a neo-liberal turn that generates the conditions for a more mobile self but, as Powell notes, this is underpinned by a political agenda that has at its core "the marketization of everyday life" (Powell, 2013: 1). Elliott and Urry comment: 'The advent of digital technologies involves the creation of new kinds of mobile life, new kinds of sociality and new ways of relating to the self and others' (2010: 45). In effect, this can position the individual not only as consumer but also as 'the product'. It is therefore no surprise that the kind of PR, marketing and language once applied to products and services have been employed to projects that package the self in particular ways for the market place. As Elliott and Urry have commented, this shift is not unproblematic and if anything can create more inequalities between those who have economic, cultural and *network capital* and those who are excluded. This is worth further consideration as promotion through social media is not just about having a visible presence but, at a deeper level, about accruing network capital. Elliott and Urry conclude by considering that a "good society" should be a "socially inclusive society" and that network capital 'should be enlarged and social exclusion would be lessened through spreading such capital as equally as possible' (2010: 63).

Therefore, critically thinking about how we use digital technologies and networks and what we communicate matters.

Creativity and commerce have always had an uncomfortable alliance. Many myths have sprung up around the role of the artist and some of these have masked practices that are socially divisive. As we consider cultural and networked capital we also need to think about how cultural production is positioned in these debates. The genius artist starving in a garret for the love of art has been a romantic figure popularised from the Renaissance through to Victorian Britain. This view has positioned art and the role of the artist as distinct from the rest of society and artistic practice as a special and unique kind of labour. Janet Wolff, writing three decades ago in the 1980s, attempted to dispel some of the myths surrounding art and creativity by taking a distinctive, sociological perspective that art was situated as a social product and could be analysed in terms of the social context of its production. She argued in the book: *The Social Production of Art*, it was through particular institutional practices that art and artists came into being and through a system of gate-keeping that it became elitist and exclusive. It is worth considering how this might be understood in relation to the previous discussions of gallery spaces and the film festival circuit. Are there parallels to be drawn here? Her argument is that creative practice, like any other form of labour, is not unaffected by the social and economic relationships of a capitalist system of production (1981). Since the 1980s of Wolff's book, access to the Internet and social networking on a global scale have shifted some of the ways we think about creativity, the artist and the promotion of media work (Powell, 2013). This doesn't exclude the idea of how art or indeed labour, are positioned. Gate-keepers do exist and the gallery/dealer system breeds exclusivity. However, the Internet means that aspiring artists have a method of distribution unavailable in previous eras. The ability to bypass the more traditional mechanisms of media distribution has been used to herald the Internet as a new tool in the democratisation of a range of creative practices from art to journalism. Whilst this view is positive in celebrating empowerment and participatory culture, Powell reminds us of the need to be cautious in making such claims and that in the convergence and integration of consumption, commerce and culture on the web, forms of exploitation go unrecognised (*ibid.*).

We can argue that creativity should not be seen as separate or exclusive but within the reach of all. Part of thinking through creative mobile media is to foreground, through non-specialist technology, what can be possible with the tools that many ordinary people have access to and how using networked technology becomes a more social and participatory form of art or creative practice. That said the need to generate income and earn a living has to be recognised. Developing skills and building up expertise through time and experience is also part of the creative process. Creative practice can be viewed like any other form of labour requiring time and effort to produce an artefact that adds value and meaning.

Selling work may be one of the ways you can cover costs and support yourself. However, how you price your work isn't an easy decision. You should consider the time, effort, materials and experience that go into the work and then decide what an appropriate remuneration for this might be. You also need to consider how art dealers use exclusivity and scarcity to add value in the selling process. It is in the interests of the dealer for you to be both successful and exclusive as that garners the best price from which they may take a substantial commission. The question you need to ask yourself is whether you want the highest price you can get for, say, a photograph or whether you want to produce work that is affordable to ordinary people? Some artists do both. Making a catalogue that includes both photographs and written content can be a way of extending the work you create and/or exhibit to a larger audience at an affordable price. Setting up a free online catalogue may allow a global audience to connect with the work across time and space while individual prints are sold off from the exhibition. Consideration of your audience and what you do with your creative projects can be both a financial and an ethical decision. Part of your strategy for promotion will include knowledge of some of the issues raised around commerce and art along with your own position on the subject so you are able to make more informed choices as you proceed.

## (b) *The Artist's Statement*

The *Artist's Statement* is perhaps the first thing to consider on the road to promoting yourself and your work. The first gallery visit should have made you familiar with some of the ways artists and practitioners introduce themselves to their audience. There is no single way of writing a statement but it needs to be written in such a way that it engages people. The tone you adopt should enable your audience to understand something about you and your work. Some artists like to focus on more personal aspects of their lives; short biographical details based on their home life and education and the start of their career before focussing on the rationale behind the work on display. Other statements are less personal and concentrate more on the themes and 'big' ideas that link practice.

New artists might find it useful to include a photograph so that people can put a face to the name so it becomes an aide memoire for the visitor to the gallery space. Try to make the photograph you use say something about your style of work rather than simply using a mug shot. There is often a photograph of the author on the cover of a book he or she has written, taken usually in a study and surrounded by books. This isn't accidental or arbitrary but absolutely part of the message they are trying to communicate about themselves; the start of 'branding the self'. In this book, you will find biographies of myself and the other contributors. This is to give you, the reader, some idea of the credentials we have as academics and practitioners for writing this book and helps validate its contents. An artist's statement does a similar job and should offer an introduction to your interests and the themes and rationale behind your current practice.

## (c) *The Synopsis*

The second piece of formal writing you will need to consider, after providing the more general overview in your artist's statement, concerns the specifics of the exhibited work. You can think of this in terms of a *synopsis*. Learning to write an effective synopsis is a key skill and has numerous functions in terms of promoting your project. It is one of the first things potential funding donors will turn to for a sense of what it is you are intending. Synopses are often to be found on the walls of established galleries as you enter each exhibition room. In smaller pop-up venues this could take the form of a simple photocopied sheet that is either handed out to visitors or left in a prominent place for them to help themselves. Public film screenings often favour the handout at the entrance to the theatre. Online screenings will feature a short synopsis before the play button.

So, the synopsis can be summarised as an important and multifunctional piece of writing but what exactly does this mean? First and foremost it is a concise piece of writing that outlines for an audience what the media work is about. This can be extended to include some of the specific themes the work draws on. You can think of the synopsis as employing both descriptive and interpretive forms of writing that you encountered in Chapter 2. You may start with a short description of the work followed by your interpretation of its significance in wider cultural terms. The tone of a synopsis is key to how successful or not the text is. It is written for an audience so needs to be both accessible and engaging in order to communicate to as many people as possible. Having read the synopsis the audience will hopefully be keener still to see the exhibition or watch the film.

If you are working within particular, specialist areas the language you normally employ can be quite rarefied. Although it might be familiar to those in the same knowledge circles it may well exclude others. It can be quite a difficult balance to strike but it should not be impossible to make the text more accessible without oversimplification. Most audiences come to exhibitions because they want to learn something so are not unreceptive to new concepts or difficult ideas. However, we have all had experiences of either being put off or made to feel stupid by incomprehensible or jargon-heavy language. Communication in such cases was clearly not effective. Time spent researching how others write is never wasted and drawing on gallery visits, reading published exhibition catalogues and art and screen magazines will help you identify what works and what does not. It will help familiarise you with the kind of language commonly employed in artistic circles. A final note on writing a synopsis is that drafting and proofreading anything public facing needs to be second nature.

Getting publicity for spelling errors rather than the works in your exhibition is not the best form of PR.

## (d)  *Posters, Fliers and Exhibition Design Labels*

Branding needs both consistency and visibility and this applies equally to the way you promote or package your product. The poster and flier, together with any personal invitations, are extremely useful tools for generating publicity, creating a buzz about the work and getting information out to the widest possible audience. You want people to watch your film or visit your exhibition and to do so they need to know that an event is taking place.

A mailing list is as good a way as any to start thinking about who might want to see your work and how this could aid the generation of further publicity. Duffin (1991) calls for three distinct groups to be considered: the media; galleries, art organisations and other art colleagues; friends and family you know and would give support. You might start to think about creating a specific look across the whole range of promotional material that includes personal invitations and which can be part of an email distribution to the various groups on your list. Would your audience connect the poster they have seen in the street, the flier handed to them and the design labels in your gallery space as being part of the same exhibition? Designing your marketing materials with a view to creating this connection can be achieved through typography, colour and imagery along with any title of the work or name of the artist.

In designing a poster both size and the location — where it will be displayed — are the first considerations. Some details on the poster might need to be visible at a distance. You can begin by sketching out ideas, brainstorming or writing a word list based around what it is you are promoting. Location will also determine whether the poster works best in landscape or portrait format. As part of the layout and design there are key pieces of information that need to be included. You can think of this firstly in terms of the title or headline; the big details such as time, when and where. Then consider the smaller details, which may include web addresses, graphics, logo and any further information relevant to the type of poster you are creating. Once you have mapped out the essential content of the poster you can begin to consider the design in more detail. If you are promoting a film then its genre may give you ideas of how the elements can be organised and so create a specific style for your poster.

Typography is an important consideration and many of the most effective posters have been made through using only colour and type; not all posters need illustrations or photographs. Think about how you use space for creative effect to ensure there is a clear focus on the key message. If there is too much going on, the message can be obscured or even lost. Colour can be used creatively as part of the overall aesthetic. If in doubt keep things simple. A scaled-down version of the poster may make a good flier so giving you brand consistency across the range of promotional material. An example of a student poster can be seen in Fig. 3.3.

Creating an interactive poster may be more time consuming but you can draw in a media-savvy audience with this technique. Not only will they be able to get further information about your work through the links you create in the poster, they are also likely to be part of a generation that are active on social media and can *spread* your message. The *Blippar* app available as a free down load allows extra content to be revealed via a mobile phone as a form of augmented reality advertising. Blippar term this "visual marketing" (https://blippar.com/en/, accessed May 2016). The following section extends the idea of including an online presence to any promotional campaign and how social media can play a role in communicating your ideas.

## (e) *Spreadable Media*

'If it doesn't spread, it's dead' (Jenkins *et al.*, 2013: 1).

Jenkins *et al.* argue that the Internet has created space for a more participatory form of culture that has reframed our relationship with both media and consumption. The networked environment has shifted the emphasis away from a simple model of production, reception and

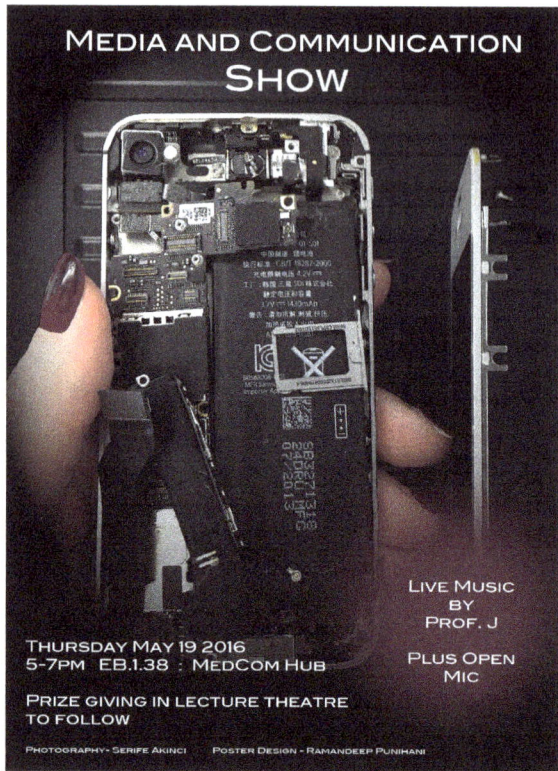

Fig. 3.3   Media show poster by Serife Akinci and Ramandeep Punhani.

distribution to consider the multifarious ways information and ideas are circulated. They use the term *spreadable media* to consider how new technology, including social media, offers platforms for people to pass on media artefacts and share and circulate content. In turn they consider the revenue generating potential through the activities of such users (2013). Spreadability is nothing new and we can consider older practices such as word-of-mouth that have always been part of the toolbox for generating publicity for an event or a film screening or persuading people to buy a product or service. Digital media does not replace this but reconfigures it for the online community. *Viral marketing* has used the metaphor of the virus, something that spreads rapidly at speed across the population. Thus, a *viral marketing loop* in the world of advertising is seen as something desirable. What this means is that if a product or service is created with *virality* people want to use it. It spreads through users, in turn generating new users (Penenberg, 2009). Jenkins *et al.* argue that by using the term viral there is an implicit suggestion the spread of a message or idea can occur without the users knowledge or consent. That is, they are seen as passive users. The term spreadable is preferred as it recognises a participatory culture and avoids metaphors of "infection" or "contamination" in which the user may appear to have little agency (2013: 21). Spreadability implies active participants who gain something through the process of spreading content. The terms *viral* and *spreadable* are either blurred or conflated in popular usage. As the media landscape evolves there will, no doubt, emerge new metaphors and terminology to describe the varied and nuanced ways people communicate and spread content and how the worlds of commerce, advertising and marketing utilise this.

We need to think about how some of the knowledge gained from promotion and marketing research might form a useful addition to ways of promoting ourselves as artists and practitioners. David Scott in *The New Rules of Marketing and PR* speaks about building relationships with people rather than the 'old idea of advertising to them' (2010: 4). Making our work visible and distributing and circulating information and publicity need strategies that embrace a digital culture. What the ingredients are that makes something spreadable are questions that marketing departments constantly grapple with. Whilst we can start to analyse some of the qualities that have produced successful marketing campaigns, or try and unpick internet *memes* and *virals,* there is no way of knowing if what we produce will ultimately be successful and spread. Internet users are at their most creative when professional marketing has been turned on its head by the appropriations and additions of user-generated content. Such users adapt content in surprising and novel ways. In *Creative Mobile Media* there is an assumption that students will embrace these new and emerging forms and contribute to the changing landscape of creative practice and promotion of ideas. In addition to the projects you create for exhibition and the traditional media you produce for publicity such as the poster, try to create a spreadable media ad that promotes yourself and your work. Think carefully about the content you produce and how you can actively engage others with it in a meaningful way. Spreadable media implies a generosity of spirit that may be lacking in some of the more commercial advertising created by professionals who need to keep control over their products or services and profits. Creating an online campaign acknowledges that as a content creator you may lose some of your control. So, for example, how would you feel if people remake or reinterpret your work on the web? Would your feelings be any different if others were to financially profit from your ideas? In creating participatory work you might consciously include a call to add or change the content. What would make it spreadable? Why would someone want *to pass it on?*

## 3.4  Keeping it Legal

The final section of this chapter draws attention to those working practices in the media and creative industries you need to engage with as a matter of routine. If you are reading this book in a specific chapter order then it may be worth you turning to the final chapter of the book that seeks to raise awareness around the ethics of working with others. This section will consider the practicalities involved. If you have people in your film or photographs we will assume you are working ethically and have considered how they are both represented in the final work and treated during the production process. To keep work legal means that you obtain the necessary permissions in the form of consent from those working in your project and that they are in a position to grant such permission. In the UK, adults are 18 years of age or older and assumed to have the mental capacity to make such a choice. Parents or legal guardians need to give consent for children and vulnerable adults. Consent should cover all aspects of the work taking into account any restrictions around exhibition and distribution that includes online platforms. There are many web resources that give examples of what permission or consent forms look like and what should be included. These range from the straightforward and readily understandable to quite complex legal documents that use specific legal terms. It is always best to use language that is unambiguous so both you and any participants are quite clear on what they are giving consent to. Figure 3.2 is a simple version of the kind of information you can expect to see on consent forms and is produced here simply for guidance, not use. If you are a student your university department will have consent forms and guidelines that you should make use of. If you are using professional actors *Equity*, the actor's union, has guidelines on contracts and information on rates of pay (https://www.equity.org.uk/resource-centre/for-employers/, accessed May 2016).

Permissions need to be obtained for any music and musical performances, not composed and performed by you, which you wish to include in your project work. The music industry has very strict copyright rules to protect composers and performers and it can be costly getting the rights for existing music to use in your project. For this reason, many students rely on *Creative Commons* licenses (see the section on Intellectual Property Rights) or seek permission from up and coming performers who may grant use of their authored works for a credit or citation in your work. Even here it is important to get the permissions in writing that clearly states whether or not you have global rights to use the music and for what length of time. Some musicians may stipulate that you have permission to use their music for, say, up to a year as long as the work has not been created to generate profit. Permissions may also be needed for many sources of sound as well as music that you find on the Internet. Whilst Internet sites are required to remove anything that breaches copyright law not all sites have been diligent in following such directives. It is quite possible to find material that could infringe copyright and using it without permission could leave you open to prosecution.

Different countries have different laws on what can and can't be photographed and it is important you find out what restrictions are in place in the country you are working in. The UK, for example, has legal restrictions both on the right to take photographs and what can be published. There are many websites giving guidelines on this but be aware that all come with disclaimers. This is a legal minefield and both The Official Secrets Act, 1911 and more recently the Terrorism

Act, 2000 and Counter-Terrorism Act, 2008 (www.parliament, accessed June 2016). have extended police powers that limit photographic practice.

Photography in public places, including the street, is usually allowed providing no obstructions are caused or results in conduct likely to cause a breach of the peace. However, difficulty can arise in determining what are private and public areas. Many places used by members of the public such as railway stations, shopping centres or local parks are actually owned privately or by the local council and where permission should be sought. For example, Trafalgar Square and Parliament Square prohibit photography that applies to "business, profession or employment" and written permission is needed for anything other than tourist photographs (Macpherson, 2004). The advice to students is to check whether they are on public or private property and to always seek written permission if in any doubt. Whilst you may be within your rights to photograph people on the street and certainly many great images would not exist if street photography were curtailed, you need to think carefully about how the images are used if members of the public can be clearly identified. Refer to the chapter on ethics in Part Two where such matters are discussed in detail.

It is important to get great photographs for your project work and if your method is driven by spontaneity this might well include the candid shot; working ethically involves consideration of others. You may need to approach people, show them the images you have taken and explain how you intend to use them. You may well be surprised to find how many are happy to give their permission. If they object then delete the relevant images and move on. Some great students projects have been undertaken in street photography where the challenges of photographing others were met with creative solutions. Changing the camera position and using the surroundings to obscure faces might be one solution with a more dynamic end result.

### (a) *Freedom of Panorama*

Architecture, statues and artworks that are located in public places have been created by someone and under normal copyright laws, licences must be obtained to photograph them. However, Freedom of Panorama (FOP) is an agreement that allows exceptions. In 2015, a controversial plan to change laws regarding FOP, which would have serious effects on both the rights of photographers and the images already in circulation on the Internet, was overturned by members of the European Parliament (Cheesman, 2015). Even so, legislation to extend it throughout Europe failed, which means there are still variations in the rules from country to country and it is important to check what the position is for photographing architecture and buildings in the country you intend to work in.

## (b) *Intellectual Property*

The Copyright Designs and Patents law can be complex in how it is applied but it does offer protection for artists, designers and photographers (Davies, 2010). Creating original work takes time and effort on your part and you have rights regarding how the work is distributed and used and that you are recognised as its creator. This is pertinent if, for example, you are trying to earn a living from your work through selling an original work of art or charging a licensing fee for someone else to use one of your photographs. The advent of digitisation and the Internet have created particular problems for photographers and now there is always the possibility that, once uploaded to the web, work can be used without permission or appropriated in ways you might not have considered or approve of. Davies discusses common ways for helping to prevent unlawful dissemination that includes the use of thumbnails or low-resolution images or by adding a watermark to photographs (*ibid.*). Of course, this doesn't necessarily prevent them from being used by others but a watermark does enable you to add copyright details and signal ownership.

Copyright law applies to mobile media in the same way it does to any other film, photography or sound project that might have been created using professional level technology. As it has been stated previously, if it is original work and you are the creator you have rights that are protected by copyright law. You need to consider the kind of protection to put in place to safeguard your rights and seek legal advice if you feel your copyright has been infringed. However, once you upload images or films onto the Internet it is not unreasonable to expect some of the people who view them to share them with others. If you find your work used by someone else without your permission, particularly if it is for commercial gain, this is very different matter and you may well have a case to bring under copyright law. The Internet has been somewhat of grey area in legal terms so it is important you check carefully the terms and conditions of using, for example, a social media or hosting sites and note any control you are asked to sign away. Participating in social media where the rationale is all about sharing can mean it is harder to prove ownership and legal cases can prove costly. It has to be said too that many people working with mobile media do so for the sharing and collaborative communities they are part of. Foregoing some rights or being generous towards those who may seek to use their work on the Internet is a choice some creators consciously make for the benefits of being part of a wider creative community. *Creative Commons* and *Copyleft* are both organisations that have been developed for the self-management of intellectual copyright and to make public work that can be shared with others. For further details see http://www.creativecommons.org.uk/ (accessed May 2016) and https://copyleft.org/ (accessed June 2016).

## 3.5 Conclusion

This chapter concludes the practical guide of Part One by considering how the projects you have created can be promoted and shared with others. There are a wealth of opportunities to join like-minded communities for the sharing and exchanging of ideas and information and in doing so, forge both a new mobile aesthetic and practice, and create new knowledge about the world. The Internet has made it easier to have a presence not bounded by time or place and for creators to bypass the more traditional cultural institutions that can limit access and visibility. As a creator of original work you have recognised an ethical dimension to working practices that is relevant to how you treat others. It is recognised too that, as a creator, you have rights and copyright law, although complex, exists to protect the exploitation and appropriation of your work without permission. The Internet has presented many challenges in this respect and whilst artists continue to look for ways to protect their work and livelihood, an online platform has brought with it many creative freedoms. Demonstrated through practice, whether it be photography, blogging, film or music, a creative spirit flourishes and extends to the building of communities. *Creative Commons* and *Copyleft* embody this generosity of spirit but so too do many others working specifically in mobile media. The mobile phone has proved to be an easy, relatively cheap and accessible creative tool that has the advantage of connectivity making it possible to share creative projects with others.

You may have already engaged with some of the chapters in Part Two of the book but if not, it is time to read how key ideas have shaped both thinking and practice. The theoretical and the practical are, for *Creative Mobile Media,* integrated rather than discrete categories. Through extending reading and research and by being actively engaged in debate and critical thinking we can broaden our understanding of the world and the people who inhabit it. Practical projects are part of this research landscape and contribute to ways of understanding that can impact on the lives of others in our community and beyond.

# Part Two

# Chapter 4

# Always On: Mobile Culture and Its Temporal Consequences

*Helen Powell*

'The jinn are poor judges of duration. In the jinn world time does not so much pass as remain. It is human beings who are prisoners of clocks, their time being painfully short.' (Rushdie, 2015: 194)

Television advertising has the potential to be positioned as a derivative of anthropology, combining a visual emphasis on the human condition and its fundamental relationship to goods and services. If we study advertisements from any decade, the scenarios depicted signify problems and challenges faced by consumers in everyday life and how choices, in one brand's favour over another, of course, allow for resolution. However, a recent ad for the Lexus NX F Sport (YouTube, 2016) introduces an alternative problem-based scenario: one in which the brand itself must seek resolution. The ad opens with a close-up shot of a female hand, texting on a smartphone which then opens out to a ubiquitous city street scene where the commuting masses are walking in all directions simultaneously glued to their mobiles. In the next shot we are in a restaurant where couples are on their mobiles whilst dining and then inside a boutique where the clothes being selected gain only a cursory glance compared with the consumers' fixation with the small screen. What does this advertisement tell us about car production and consumption? As the tagline indicates, as "the striking Lexus NX" enters the frame, so it denotes eye-catching design whilst connoting that brands have to work extremely hard to be noticed in an eyes-down mobile culture.

And yet this is not the fundamental problem here, as the rise of mobile advertising would seemingly testify, with many brands using smartphone technologies to their advantage. Rather, this chapter engages with a more significant issue, one that examines why have we become so dependent on our mobile phones? The chapter will seek to deconstruct this question through three inter-related concepts that will form its structure: uncertainty, risk and temporality. Through examining our historical relationship to time and timepieces, coupled with what Beck has defined as the rise of a 'risk society' (1992) we will navigate our way through the development of a culture of mobile dependency. Utilising, but also challenging, Elliott and Urry's (2010) concept of 'networked individualism', this chapter analyses to what extent we develop personalised, nuanced, multilayered temporal registers to get us through the day? Or, could it be that instead of individualised approaches to time, what we see in mobile media is a return to more consolidated and repeatable homogenised behaviours and patterns emerging in relation to our diurnal perceptions and practices? Driven by conceptions of time as a quantifiable resource, are we witnessing the creative potential of the individual's relationship to time beginning to be crushed in this current era of disorganised capitalism as it was with the introduction of the clock in the first industrial revolution? And is this due to the wider consequences of living in an age of uncertainty permeated by risk? Alternatively, might we proffer a more positive reading which positions the mobile as a fundamental conduit for reassurance, connectivity and indeed meaning making throughout our daily lives and above all, as a tool of temporal control and navigation?

## 4.1  Signs of the Times: Uncertainty and Risk

The political events in Great Britain across the months of June and July 2016 were nothing short of extraordinary. A referendum and a marginal 'leave' vote was followed by Brexit that then led to one Prime Minister resigning and another moving into Number 10 before the incumbents had even finished packing. The neighbours on the other side of the House were also in a state of disarray and, at the time of writing, the Labour Party is teetering on the brink of division with the potential for this rift to open up into a permanent split. Meanwhile, across the Channel, Bastille Day was plunged into chaos as a single terrorist caused carnage along the Promenade des Anglais in Nice. The country entered 3 days of mourning and remained on high alert. How can we describe this state of affairs, one characterised by 'unknown unknowns', and how does this relate to mobile phone culture beyond the significant role it plays in capturing and communicating events almost in real time?

The series of events described span just three weeks, from June 23 to July 14. That so much should happen that was largely unpredictable is simply beyond belief. However, it could be argued that we live in an age predicated on uncertainty. We have never been able to control the future, although the 'grand narrative' of progress that underpinned modern thinking worked towards that goal. Its fragility was always reflected in the culture of film and literature, especially science fiction, and it is interesting now that the cultural barometer swings away from the conquering of space to apocalyptic and post-apocalyptic visions. The fault line of uncertainty that resides beneath our feet 'appears at unexpected moments…the cunning of uncertainty excels in uncovering the unintended consequences of human purposeful action' (Nowotny, 2016: x). As a result there is often a misalignment between our intentions and their outcomes. But what is the difference between uncertainty and risk?

As noted, one of the principle legacies of modern science was the ability to predict and therefore control the future on both a macro and micro scale. On the one hand, the rise of online betting carried out on your phone through companies such as Bet365, is testament to man's fascination in the ability to anticipate outcomes. Pollsters, such as YouGov, gain their credibility through forecasting outcomes, often of voting behaviours: predicting in June 2016 that Britain would 'Remain' by a small margin (YouGov, 2016). However, the rise of analytics and programmatics in advertising and the use of Big Data more widely, speaks to an age of increasing consumer unpredictability. The late sociologist Ulrich Beck coined the term 'risk society' back in 1986 and published a book of that name in 1992 developing his ideas further in 1994. Reflecting on the rise of 'risk management' and risk assessment' as increasingly integral parts of the working life, late modernity was seen as being underpinned by a recognition that with every decision comes some kind of risk. Beck (1994) distinguishes between a 'pre-industrial society' (traditional society), an 'industrial society' (first modernity) and 'risk society' (second modernity). In each epoch so the nature of risk changes from natural hazards, such as floods, to manufactured risks such as pollution and oil spills. These can happen anywhere at any time and the consequences can also impact over much longer periods of time. Therefore, 'risk society' emerges as a consequence of 'industrial society's unwanted side-effects' (Heaphy, 2007: 80). Or as Beck defines it:

> A phase of development of modern society in which the social, political, ecological and individual risks created by the momentum of innovation increasingly allude the control and protective institutions of industrial society. (1994: 27)

We consistently observe structural changes in relation to working patterns, family life and indeed social relationships and these then impact on the very nature of risk itself. In essence, this comprises two dimensions that are inextricably linked: as society becomes more fragmented and individualised so we spend more time negotiating choice. However, as a result, individuals must assume a greater degree of responsibility for the consequences of these actions. In this era of 'reflexive modernity' (Beck, 1994: 2) self-threats dominate that are far more difficult to control. These then are amplified through social media: we become more aware of the levels of change on a daily basis through the immediacy of news, and thus we amend our behaviours accordingly, producing further change at the micro level. Therefore, we might argue, 'the difference between risk and uncertainty is whether the possible outcomes can be calculated before or not' (Nowotny, 2016: 70).

## 4.2  The Role of the Timepiece in Modernity

Consideration of time and temporal experience came to the fore with the advent of modernity and in particular industrialisation, which heralded a series of significant changes in terms of how time and space were conceived. Accompanying the standardisation of time across the railway transportation network during the 1870s came the recognition for the need to find a universal measure to co-ordinate working patterns with the spread of industrialisation. Clock time was adopted as a functional standard in that it could be adapted to any mode of work. 'Fundamentally its appeal lay in the fact that it engendered precision whilst denying interpretation' (Powell, 2012: 115). In Thompson's ([1967] 1991) *Time, Work-Discipline and Industrial Capitalism* he argues that the current Western approach to time, quantitative in nature, was fuelled by the Industrial Revolution which saw a shift from agrarian to factory-based industry. Inextricable to this was a reconceptualisation of the perception of work's relationship to temporality, from one in which tasks to be completed is replaced by an idea of 'time as currency'. Thompson (1991: 354) commences his analysis by problematising the impact of the appearance of the clock: 'how far, and in what ways, did this shift in time-sense affect labour discipline, and how far did it influence the inward apprehension of time of working people?' As people began to work for others, as opposed to working for themselves, 'time-discipline' became a requirement (Thompson, 1991: 388).

In this seminal essay on time Thompson (1991) documented how our relationship to work was fundamentally changed with the introduction of the clock. Instead of the completion of tasks being based on available light, the introduction of the factory system, made efficient through the findings and implementation of time and motion studies, saw productivity measured on the basis of the completion of a certain amount of work per hour (Powell, 2012: 116). One of the legacies of the industrial revolution and cornerstone of the functioning of modernity was the evocative symbol of the pocket watch presented on retirement and representative of the time given to the company in question. Or potentially that retirement is all about being given one's own time to do anything but work: and in effect, allowing you to step outside the constraints of a rationalised and uniform approach to time that served the rise and maintenance of capitalist production. Industrialisation was, therefore, to a significant degree about man's control over the time of nature but in fact through the invention of clocks and watches time created to human design soon became 'naturalised' (Adam, 2003: 62).

## 4.3　Regaining Control: The Self as Time Lord and the Mobile Phone as Tardis

Let's take a moment to recap the argument so far. The condition of late modernity, or what some theorists have called postmodernity (Harvey, 1990) has been characterised as an 'age of uncertainty' (Nowotny, 2016) or as a 'risk society' (Beck, 1992). Coupled with this, a key feature of late modern or postmodern society is its temporal complexity. Indeed, Harvey (1990) centres his understanding of postmodernity on how we experience space and time. He believes a key feature of postmodernity is 'time-space compression', namely 'processes that so revolutionise the objective qualities of space and time that we are forced to alter, sometimes in quite radical ways, how we represent the world to ourselves' (1990: 240). How we understand and experience time, therefore, is highly informed by our social surroundings and the material world and in this context, specifically the embedding of mobile technologies into our lives. The mobile phone changes how we manage and indeed understand space and time. For example, space–time compression occurs each time we take a call on our mobile phones. The mobile has come to be associated with the concept of presencing: we expect the other person to be on the end of the line ready to respond, but Moores (2004: 30) is more specific, referring to this phenomenon as 'presencing at a distance'. All too often we hear that familiar 'I'm on the train' as when we speak on the mobile, as opposed to the domestic telephone, it is not clear to the caller where we are. When we call a landline, we call a place not a person. Furthermore, mobile calls can come at any time, day or night, in any place: therefore their receivership has an impact not just on us but on anyone else who is surrounding us.

How we perceive and engage with time has reached new qualitative levels: our sensitivity towards time has become heightened and our responses to it more emotionally charged. The availability of so much information at our fingertips means we require more time to process it and this sense of time pressure is added to by the almost instantaneous experience of sending and receiving across mobile communication devices (Fig. 4.1). Paradoxically, whilst we require less time to travel, to communicate, we feel that life has speeded up but this is a causal factor of the range of opportunities available and our response in trying to pack more and more into each moment, what Florida (2002: 144) calls 'time deepening'. Living in the age of the instantaneous produces a culture of acceleration. This perception that daily life is accelerating produces time pressure and time thus takes on more of a quantitative nature, needing to be managed through devices such as the Apple Watch, explored below. Technology, however, was surely meant to give us more time but our relationship to our mobile phones makes us even time poorer. As a result, the adoption of specific strategies has arisen to allow for time management in the seemingly extended present. Florida (2002) draws our attention to one of these strategies known as 'time-deepening'. Here if we cannot elongate time then perhaps we can deepen it, intensify it. Essentially this centres on engaging with activities that take less time to complete or where more than one task can be done at the same time. An interesting example of this was noted in July 2016 whereby sales of audiobooks was seen to have risen 148% in the 5 years since 2010, brought about by people who love books 'but don't have the time to read them' (*The Week*, 2016).

Increasingly people are configuring time in their own way to suit them and drawing on virtual culture considerably to do so. This Lash and Urry (1994: 246) term 'social desynchronization' and categorise it as follows. 'There is a greatly increased variation in different people's times. They are less collectively organised and structured as mass consumption patterns are replaced by more varied and segmented patterns' (1994: 246). Thus, mobile devices produce new temporalities and spatialities reconfigured on the basis of challenging accepted binarisms: presence/absence; inside/outside; public/private; near and far. To this extent Sheller (2004: 39) suggests that the mobile now allows for the reconfiguration of traditional spatial categories with 'new modes of public-in-private and private-in-public that disrupt commonly held spatial models of these two separate spheres.' She recognises that mobile social interactions are difficult to delineate arguing for a more 'messier imagery of liquid social dynamics' (2004: 41) that corresponds to a more 'fluid' understanding of social space (*ibid*.: 46). Interestingly, Sheller rejects the more traditional language that underpins the structure of mobile communication: namely, 'networks', which she argues, are 'unable to depict these processes of **uncertainty** and dynamic social change' (*ibid*.: my emphasis). In this context of new spatial and indeed temporal relations, Sheller (2004: 49) wishes to develop a new metaphor to explore models of connectivity in the age of the mobile:

Fig. 4.1   Man at bus shelter by Helen Powell.

The communication system no longer operates like a network, but has instead taken on the characteristics of something far more like a gel or viscous fluid. Persons themselves are not simply stationary node in a network but are flexible constellations of identities on-the-move.

To master risk is to master time: namely the future. Consequently, people look to seize back control, to reconnect and the mobile phone is integral here. The mobile phone is an example of convergent media: it allows us to do many things at once. It now not only shapes our sense of time but orders our sense of being within the diurnal register, minute by minute.

## 4.4 The Metropolis and the Mobile Phone: Experiences of Time in the Age of Super-Connectivity

When Simmel, the German sociologist, wrote of the modern city (1903) it was informed by a wish to document the experiences of a new set of surroundings and the impact that the sensorial overload produced had on the modern subject. Concerned that the subject would become too emotional to function effectively, a 'rational manner' (1903/1971: 326) was required in order to protect 'the inner life against the domination of the metropolis' (*ibid.*: 326). In his essay *The Metropolis and Mental Life*, Simmel documents a number of ways which would allow for a sense of detachment, of depersonalisation to occur. In general, he addresses the development of a 'blasé attitude' which he describes as an 'adaptive phenomenon…in which the nerves reveal their final possibility of adjusting themselves in the content and the form of metropolitan life by renouncing the response to them.' (*ibid.*: 330). However, adoption of a particular sensibility is not enough: modern tools are required to support such an approach. One such 'tool' being the introduction of a 'money economy' (*ibid.*: 330) which reduces 'qualitative values to quantitative terms', that is to say the introduction of the price tag nullifies the language of bartering (*ibid.*: 328). In order to coordinate the masses in the city on a daily basis a temporal framework is required that is not subject to individual negotiation and 'externally this precision has been brought about through the general diffusion of pocket watches' (*ibid*.: 328). Interestingly, he goes on to observe that so significant is this invention, this second tool, that,

> if all the watches in Berlin suddenly went wrong in different ways even only as much as an hour, its entire economic and commercial life would be derailed for some time. (*ibid*.: 328)

If Simmel were writing *The Metropolis and Mental Life* today, it could be based around the smartphone and the invisible digital temporal architecture that it orchestrates. The visibility of clocks and timepieces at stations and in factories popularised around 1900 allowing mass synchronisation still exist but their role has become somewhat diluted as indeed within the consumer culture of the 21st century the rise in the number of 'non-places' (Augé, 2008), shopping malls, service stations, airport departure lounges with their lack of seating, depend on the negation of time for profit. To amble, to peruse, to lose track of time is all the more conducive to spending.

In an age of uncertainty one of the principle ways in which we seek to take back control is through our relationship to time. It seems that if we can somehow manage our temporal relations through a mobile timepiece we can instil some sense of order out of chaos. If we think of the developments that have occurred in the potential of the mobile phone, then these are largely to do with either filling or killing time or managing information that speaks directly to the kind of life we lead (Fig. 4.2). The success of Angry Birds and Candy Crush kills the boredom of the daily commute by simply filling time. As documented, the traditional model of time was one built on the model of linearity: of time's arrow but how may we now describe an appropriate model of time? It could be argued that through the centrality of the mobile phone as postmodern timepiece we may describe time as a möbius strip, an encircling, spiralling loop with no beginning and no end: fluid, flexible and where past, present and future inter-link. We use these devices almost simultaneously to look at photographs, check the news as it happens and make plans for the evening or weekend. In this

Fig. 4.2   Girl with a dog at station by Helen Powell.

way, we seek to control time and the way in which time is managed via the mobile phone structures our quotidian existence in new and innovative ways. Therefore, it can be argued, we use the mobile like a time machine: it is not about knowing the time now but rather through its camera, its apps and its connectivity it is about controlling time and mitigating risk: the risk of disconnection; alienation; or simply missing out.

## 4.5   Networked Individualism and the Externalisation of the Self

For Thompson (1967) time was a capitalist instrument to control the worker and enhance productivity, but in the age of the post-industrial, time-poor *prosumer* so we use mobile devices as a means to time warp but also to time fill. Extending far beyond the original capabilities of the watch, not only do these devices at the macro level allow for a global temporal reach but also inform at the micro level our sensibilities at any particular moment. How we feel in time, is shaped by an 'always on' connectivity. But what do we document? This is the era of the citizen journalist but also of the inconsequential: Instagram food porn and the rise of the selfie. They document our place in the world; minute by minute. No longer do we tell our stories of what we have done in the past tense but share them almost simultaneously as they happen, in an immediate present. There is a new dynamism in the relationship between self and time via mobile media. It is no longer enough to tell the time: but locating and evidencing the self in time. We can juxtapose this presentation of the self with adoption of early photography pre-1900: so expensive was it for the masses to have their picture taken that it was only used to mark out special occasions. Now, in this participatory culture, founded on posting, feedback and response, we take everything. There is a hunger to constantly produce images and content to feed our pages. Therefore, is every moment equal? Does this mean that our relationship between time and the photograph have become devalued? Or does it mean that we are now more aware of our mortality in this 'risk society' and want to document, hold onto, each moment on earth? As a result new technologies alter how we conduct our lives and what is deemed meaningful but this can produce a generational conflict in terms of expected behaviours. These inter-generational differences are observed in Jonathan Buckley's novel *The River is the River* (2015) which explores a relationship between two sisters Kate and Naomi. They are in a restaurant and one sister observes a girl who is part of a couple:

> '"Did she just take a picture of her food?" the sister asks. The girl is performing a complicated procedure on her phone, at speed, with great concentration. 'Now she's uploading it isn't she?' says Naomi. The salad is going online. All over the country people will soon be looking at those tomatoes ... What's the point?' 'It's all about sharing' says Kate, 'We're too old to understand.'"

In modernity time held us together, now in the age of the prostheticised self, the mobile functions as a third hand. In a tap it allows us to move between past, present and future re-enacting and remembering shared times and experiences which holds individuals together whilst reaching out to future plans that will allow for bonds to be maintained. Time, as discussed above is viewed as a commodity: a resource. Our lives are made up of time, something discernible, finite which we feel we should fill up with positive experiences. Time is a valuable commodity because we simply can't buy it, there is no temporal equivalent of quantitative easing: rather a key trend in app development is all about ways to get the most out of life, manipulating time in order to gain 'flexibility, adaptability, reflexivity' (Elliott and Urry, 2010: 4). But how does this revised relationship to time impact beyond the level of the individual to society at large? As this chapter has noted, psychosocially we have seen a turn towards a more individualised self: and yet the centrality of the mobile in our everyday lives is that it allows us to remain connected virtually as it mediates our relationships. As traditional and personable structures of connectivity become eroded, work patterns for example, so our reliance on more 'abstract systems' (Giddens, 1991) become prevalent. As Elliott and Urry suggest 'the term "mobile lives" suggests an increasingly complex, detraditionalised patterning to personal life' (2010: 11). Two examples, again taken from advertising, exemplify 'abstract systems' in our daily encounters and both relate to new forms of payment and the shift towards a cashless society. Whilst at the turn of the 20th century Simmel (1903/1971) documented the shift towards money over barter as a way to increase efficiency in our daily lives through the depersonalisation of transactions, the introduction of Apple Pay (*Metro*, 2016) and the PayPal App (*Metro*, 2016) takes this a step further. This distinction underpins the tagline of the PayPal ad: 'Old Money Talks: New Money Taps'.

In an early study of mobile phone culture, Sadie Plant (2002: 23) observed how the mobile's accessibility 'alters the way in which individuals conduct their everyday lives'. She argues that when we are carrying a mobile we are always temporally fractured or discombobulated, in two temporal moments at once: the absolute present and a potential future when the phone might ring or a text might be sent and as a result 'just the knowledge that a call might intervene tends to divert attention from those present' (*ibid.*: 30). Developing this idea further she explores how we then become 'adept at operating as though in two worlds ... in a way the mobile has created a new mode in which the human mind can operate, a kind of bi-psyche' (50). Another dimension of this bi-psyche, I would argue, is that of the absent presence. If we cannot be seen in person due to increased desynchronisation, we feel that we must be visible online: in friend-ships, to remain in the loop; professionally, to evidence our qualities and successes beyond the immediacy of our place of employment, and perhaps ahead of our next planned move. From a psychosocial per-spective this emotional labour and temporal investment can generate significant degrees of anxiety, guilt, stress and pressure as we constantly seek to construct and represent narratives of the self or rather present the self as brand. The presentation of the self becomes a series of 'games' (Bauman, 1996) where the rules are subject to constant change and hence updates are constantly required. This 'game of life' take place in a 'continuous present': it is 'fast and leaves no time to pause and think and draw elaborate designs' (Bauman, 1996: 24, 25). That which is valued has high impact at first but its appeal is short-lived, only to become obsolete. This is an era of trends and fads. Pre Snapchat, Instagram, Pinterest, Facebook and the phenomenon of 'likes', Bauman (1996: 25) leaves us with this advice:

> Above all, do not delay gratification, if you can help it. Whatever you are after, try to get it <u>now</u>, you cannot know whether the gratification you seek today will still be gratifying tomorrow.

## 4.6 Two Short Case Studies: How Do We Interpret and Manage Risk as Part of Our Diurnal Activities?

(a) *The Apple Watch*

As discussed, the watch was once the central co-ordinating tool of moder-nity, revolutionising both the transport system with the introduction of national timetables and functioning as an instrument that defined the parameters of work and leisure. Today, I would argue that for many, the traditional watch functions more as a fashion item or a totem of conspicuous consumption and therefore in this context it is interesting how the Apple Watch is marketed. When presented with one of the first print advertisements for the Apple Watch being sold at John Lewis (*Evening Standard*, 2015) I am struck by the relationship of image to text which challenges my preconceptions of what a watch should be. For under an image of the watch around a wrist it is described as follows:

> Receive and respond to notifications in an instant. Track your daily activity. Control your music using your voice. Pay for groceries just like that. With Apple Watch, important information and essential features are always just a raise of the wrist away.

Telling the time is not noted: not even temporal accuracy. The concept of 'watch' as we once knew it is now simply connoted through its wrist position. In the mobile age, what we understand by the watch becomes something contested, more fluid and complex than simply the circulating hands or the digital numbers of the early Casio of my youth. Propelled by a turn to immediacy and risk management, the Apple Watch is all about efficiency and control. This stands in opposition to more traditional timepieces, anthropomorphised through hands and faces that delineated that somehow time was being controlled elsewhere, that they were working under someone or something else's steer. Now the wearer, or should it be user, of the watch is in control.

As we look at the 'face' of the Apple Watch the clock still remains central but it is small in the context of its surroundings: a solar system of icons orbiting time. In one review it was noted that 'it's a distinctive piece of tech that at first can be hard to get your head around, or even to quite know what it does' (Phelan, 2015). Requiring an iPhone to function, it sends notifications and other information to your wrist and back again, although it does tell the time without the proximity of the phone. The icons can be selected and arranged depending on your needs and what is most important for you to know in the glance of an eye, and this is not just about the time. We might argue that 'the glance' is the primary visual tool and in the context of mobiles, these glances all add up: 'collectively UK consumers check their smartphones over a billion times a day' (Deloitte, 2015). Positioned as Apple's most personal gadget to date, it speaks to the cash-rich, time-poor, interacting with their existing range of devices and entangling them even further within an 'always on' culture. As a result of these mobile devices, our relationship to time changes fundamentally:

> Postmodern iTime is not boundaried in the way that modernist clock time is, not to mention pre-modern natural time (based on sunrise and sunset). During iTime, people communicate asynchronously, through saved messages. And they are available at all hours of the day and night, incorporating their smartphones prosthetically. (Agger, 2011: 132)

I am at Athens airport with a colleague from another university. As we queue to check in, I notice the Apple Watch he is wearing and he talks me through its features. When I ask about the watch element, as timepiece, he shows me how you can customise the face. As we proceed to the desk, he uses the watch to check-in. I feel like a dinosaur with my printed digital boarding pass and ticket.

## (b) *eBay: À La Recherché Du Temps Perdu*

In the early years of the Internet it was predicted that its impact on retail culture would be significant. That we would see the 'Death of the High Street' circulated across the trade press and it would be only a matter of time before we would be shopping and banking in our pyjamas. The experiential and social dimensions of retail culture were very much underplayed and prophecies of the replacement of bricks by clicks were significantly exaggerated but nevertheless the rise of online retail (e-commerce) is a fundamental element of change within our social experiences. Retail apps have certainly changed the patterns of consumer behaviour: we can shop from our phones thus disembedding consumption from the specificities of place and space.

Industrialisation, as discussed, brought about a more synchronised approach to time and in terms of retail culture standardised opening times for stores followed, including closure on Wednesday afternoons in many places and Sunday trading is indeed a relatively recent phenomena (August 28, 1994). The relationship between consumption, space and time becomes disaggregated with the introduction of the mobile, and in the context of this case study, specifically with the rise of eBay.

eBay is a digital marketplace bringing buyers and sellers together with transactions taking place via auction, offers or through 'Buy it Now' options. Beyond the realm of the general public, many businesses, local and global, now use eBay as a virtual shop window thus extending the range of choice available in searching for any item. eBay's success speaks to the era in which we live: consumer tastes fuelled by living in a time of austerity, tinged by a hint of nostalgia, fuelled by an ethical stance towards recycling and upcycling and driven by the power of fast fashion and the need to keep up to date. In this context, how the goods are presented in terms of both image and text are critical to the success of the sale. It is not the presentational style of the platform that is of significant interest here, but rather how eBay produces a particular relationship to time for the bidder once a transaction is entered into and as such evidences the complex digital temporal architecture that has been explored in this chapter to date. That is to say, once a bid is made on eBay an alternative temporal perspective emerges.

Many items can be bought for a named price, others are open to offers but the majority use the principles of gamification to drive transactions. As an online auction site, each item sold will have a description, an opening bid price and fixed period of time across which bids can be made. A series of tactics then arise: bid early in the hope that no one else will have an interest in this item or to wait and see how the initial price rises. eBay itself accepts and publicises that most bids take place in the last 2 hours. Whatever strategy the bidder wishes to adopt a new temporal register will emerge: countdown to the end of the bid which is inextricably linked to a new temporal relationship to the mobile phone. Once interest of an item has been recorded, either by watching or actually bidding, cookies generate texts reminding you that the clock is ticking; once you bid and then are outbid, a personalised text prompts you to respond, to bid higher as time is running out. It is as though eBay does not want **you** lose out. The last 2 hours are key: all external events, conversations, encounters are framed within the context of how you are managing your bidding. The adrenalin kicks in and all else must be put on hold. You wait, you wait, we are now down to the final seconds, hold your nerve, make that final bid. 'Congratulations! You have won this item'. This is not a transaction it is a game. And if you have not won, imme-diately nudge tactics are applied: similar items are put on display to take you back into the site.

eBay changes the whole dynamic of the established relationship between the consumer and the material object. In a traditional retail context, there are mass consumers and mass numbers of products to buy but eBay's appeal is generated by a singularity coupled with a finite window of temporal opportunity. With such scarce resources at play it is perhaps no wonder that online tools have been developed to bend time's arrow. Sniper apps such as Baytomat and Myibidder function on the principles of automated bidding for a fee via either a piece of software or from a website service that place a bid with seconds to go until the end of the auction. This ability to control the game in this way may generate success for its users but the emotional distress for those who lose out is significant. It is not so much that the item is lost, that is just one part of the disappointment. More signifi-cantly time has been lost that cannot be recovered in terms of the degree of temporal investment in following the bid's progress over the past perhaps eight or nine days. In essence, sniper apps function as a thief of time: it allows us to short-circuit the waiting game. For others, they represent lost time that can never be reclaimed.

## 4.7 Conclusion

We live in an age of 'social desynchronization' (Lash and Urry, 1994: 246). In the West, people are less collectively organised and instead 'gel' (Sheller, 2004) together via individualised temporalities that are nonlinear in form. A digital temporal architecture now holds us in its grip mediated by the mobile phone. Digital technologies produce new temporalities. We have seen the externalisation of the self-evolve, a new prosthetic culture where we define our self through our objects, specifically in this case, highly personal and personalised mobile phones. This 'always on' culture negates distance as an obstacle to prevent us from remaining in touch but more significantly, I would argue, it allows us to control our relationship to that priceless commodity, time, and in so doing manage the risks that time holds in its grasp. The mobile in the 21st century earns its centrality in our lives as the principle conduit through which we mediate the world and remain connected to others: a conduit of sociability, irrespective of temporal and spatial divides. However, it is also a device of distraction, of diversion: filling our down time through the multiplicity of entertainment apps available for download.

Through mobile devices time 'is now beginning to appear as many things nestled closely together' (Moran, 2013: 16) and in this respect it is possible to argue that whilst the watch or clock symbolised modernity the mobile phone is symbolic of the postmodern or late modernity. Times, not time, flow through it. To be without one's phone is seemingly to be outside of time itself. If we live in a culture that is 'always on' then 'the most fascinating aspect of the adoption of the smartphone is the extent to which it has become not just our primary access to digital content, but an ever more comprehensive and capable remote control for life' (Deloitte, 2015).

# Chapter 5

# May Days: An Examination of Mobile Filmmaking, Family and Memory

'Modern societies need to create "sites of memory" because without them events commemorated might be effaced from recollection.' (Burgin, 2004: 21)

'Without our memories we would be lost to ourselves.' (Fernyhough, 2012: 5)

*May Days* (2012) is a short film exploring memory, belonging and autobiography. Mobile phone technology is utilised in the film as a recordkeeping device but is also implicated in the formation of memory for one Alzheimer's sufferer.

May, my mother, lived in a care home in the New Forest, Hampshire, UK. I found that on my fortnightly visits I turned to using a mobile phone camera to record our days out together. Ordinary cameras proved to be too intrusive to someone who really never enjoyed having her photograph taken. It also allowed for more candid filming of our trips to cafes, the beach, the forest, supermarkets and the pub. The clips could be played back several times over the course of each trip and discussed. In many respects, they became a stand-in for the short-term memory that May, through Alzheimer's, no longer had. The clips became the foundation for a film documenting days out with May before her death in 2014 and the mobile phone became not only a recording device but also a therapeutic tool for both of us.

This chapter will examine, from a personal perspective, how it is possible to use mobile phone technology in unexpected ways: that personal family archives may have the potential for a wider social benefit. Through a discussion and analysis of the making of *May Days* the chapter continues with an examination of Burgin's notion of "sites of memory" (2004) and with particular reference to Alzheimer's disease, Marc Augé's (2004) idea that memory "needs forgetfulness."

## 5.1  Context

I have always been interested in how new technology gets used both in everyday life but also in creative practice. So it is not surprising that, as a practitioner, I began to use my mobile phone camera to record a series of family snaps for what was initially a personal album. In 2001, my mother, May was diagnosed with a form of dementia commonly known as Alzheimer's disease and named after the German neurologist Alois Alzheimer. According to the British Alzheimer's Society it affects around 520,000 people in the UK today and is the most common cause of dementia. Much research is currently being undertaken in the condition but at present it is thought that proteins, beta-amyloid fragments, interrupt signals between synapses preventing information from being passed from cell to cell. Protein *plaques* and *tangles* develop in the structure of the brain, leading to the death of brain cells. The resulting symptoms can include loss of memory, mood changes and problems with communication and reasoning (http://alzheimers.org.uk, accessed November 2015). Not confined to the aged, early onset Alzheimer's can affect people in their 30s and 40s, however, it is most likely to affect a much older demographic that included my mother. Alzheimer's is a progressive disease and as yet there is no cure.

No longer able to live at home May eventually went to live in a residential care home in the New Forest, Hampshire, UK. The area is a designated National Park and an area of outstanding natural beauty. The landscape takes in forestry, heathland and is close to the beaches of the south coast of England. This natural landscape became a backdrop to the film *May Days*. Every two weeks I would make the journey from London to the New Forest to visit my mother. I would take her from the care home and we would drive around the countryside stopping on the way for lunch and chatting about our day. I turned to using a mobile phone camera rather than any other device to record our days out together. My mobile phone was always with me and as a small, portable device allowed for discrete filming in cafés, shops, beach and forest. The clips could be played back several times over the course of each trip and their contents discussed. The material acted as an aide memoir and became the starting point for conversation.

My research led me to uncover studies where cameras had been used to aid autobiographical memory in neuropsychological rehabilitation experiments. Recent developments in memory studies, including the use of neuro-imaging scanners to detect active parts of the brain, have resulted in dramatic changes to how we understand memory. It is now believed not to be a single, unitary process but composed of a variety of distinct processes and systems. Memory is dependent on signal transmissions from billions of neurons in the brain and specific parts of the brain contribute to different memory processes. We also now have new knowledge on how memories may be stored and retrieved but rather than composing of literal recordings of reality they are in fact shaped by our present (Schacter, 1996: 5). The term *memory* therefore covers the distinctions between stored sensory and emotional information, semantic memory (memory for factual information) and episodic memory (memory for events). It includes long-term and short-term memory, the ability to recall events from the past and events that may have happened a minute ago. It involves too, non-conscious memory that underlies tasks such as playing a musical instrument. In recalling the memory of our own lives (autobiographical memory) we draw on the different kinds of memory and integrate them in to something that gives us meaning shaped by our beliefs and experience. Fernyhough states: 'Memories are changed by the very process of reconstructing them' (2012: 13). The fragile nature of particular kinds of memory with its subjective quality and its reliance on repetition, to create a memory trace in the brain, is thus far removed from the idea of simple storage and filing systems where data can easily be retrieved.

How Alzheimer's symptoms first manifest themselves in an individual is dependent, particularly in the early stages, on what parts of the brain are affected. It is not uncommon for families that care for a member of the family with Alzheimer's to be distressed by behavioural changes in that person, which seems completely out of character. My own mother did change when I look back across the period from the onset of the illness to the time of her death in April, 2014 but, as her family we were grateful that we never had to experience some of the more severe changes in character that others have documented. She kept her sense of humour, stoicism and with the occasional prompt, remembered the names of her immediate family and the fact that she had a son, a daughter and two grandchildren.

The question I ask myself is at what stage did I move from taking family snaps for the personal archive to documentation, then film for a more public audience? This came gradually. As a practitioner I was always conscious of the properties of photography as a medium; to *fix* things in time, to preserve *a likeness*. My documenting began as a rather desperate attempt, as a daughter, to capture the mother I knew before she slipped away and became lost forever. I wanted to freeze her in the frame and preserve her from the inevitable changes that were to come. However, the documentation of May's progression through the disease gradually transformed my own understanding and acceptance. I found that on my fortnightly visits to see May in the care home using a mobile phone camera rather than any other device to record our days out together forged new memories. These became centred on my mother in the present and as she was rather than the mother I had created in my head or the mother I wanted her to be.

Recording our time together worked to enhance her individual experience of Alzheimer's in a positive way. She would surprise me too when a new short-term memory lingered, a brain pathway re-connecting for a moment to produce a confirmation: 'Haven't we been to this cafe before? Isn't this place Milford?' I am convinced the constant repetition of our visits and activities had some effect on May to build up a sense of place and at the least make her feel comfortable in the environments she somehow sensed she had been to before.

However, there was nothing in the way of a controlled study to provide evidence in the medical sense as this was not the focus of my work.

Fernyhough makes references to a study on how a SenseCam was used as an aid to autobiographical memory as a form of neuropsychological rehabilitation in patients with forms of amnesia. The SenseCam, a small digital camera that was hung round the neck with a lanyard, takes pictures at fixed intervals and could record a 'visual diary' of the users day. The images can be transferred to a PC and replayed at will. Results showed benefits in some patients of reviewing the images daily and some retention in the memory of the SenseCam events. SenseCam images are recorded from the point of view of the self and Fernyhough suggests this brings them closer to the visual nature of autobiographical memory (2012: 172). The scope of research being done in the field across a range of medical conditions involving memory loss offers hope to many and MRI technology utilised to build up a complex picture of how much of the brain is active in the creating, storing and retrieval of our memories has given us a clearer picture of such processes. The repetition of the visits and the return to familiar places was a routine that established familiar patterns in May's life. Whilst I cannot make any medical claims in the treatment and care of Alzheimer's patients, in the specific case of May, the creating and sharing of visual memories of the day, I would argue, worked to enhance her individual experience of Alzheimer's in a positive way.

Photography along with the use of music has been encouraged in those who work with the elderly and in care homes in the form of reminiscence therapy and life story work. It is thought at the very least to provide a context for stimulating conversation and chatting about what photographs may represent and in turn improve relationships between carer and those who may spend the rest of their life in an institutional setting. Other studies have noted cognitive ability and improvements in well-being and mood in those with mild to moderate symptoms (Heathcote, 2009; Chiang *et al.*, 2010; Iriss.Org, 2015). Working with May on a project which involved filming, photography and sound capture of our days spent exploring the

New Forest gave us both plenty of material to stimulate conversation and to ponder the importance of time, memory and technology. The mobile phone and laptop became the digital repository of the current family album.

For my mother's generation there was familiarity with a family album; a physical collection of photographs capturing key moments, albeit edited, in family life. These would be brought out whenever any occasion, such as the arrival of visitors, warranted it. Passed around the group, the albums would be a useful starting point or stimulus for conversation. The transfer to digital files of family memories is thought to be one of the losses in a more social media-based online culture. There is no physical album to browse. The growing popularity of the creation of self-published photographic books may be recognising the need for something more permanent and the uncertainty, as yet, of the retrieval capabilities for future generations in the cloud-based storage technology. May had many family albums that archived some of the earliest photographs of her as a teenager through to the present and her life with grandchildren. I added to the personal collection of photographs, aware too of some of the studies on reminiscence therapy. When May moved to a care home, as her family, we made sure that the collection of family albums were place prominently in her room so that for as long as possible she would be able to recall she was part of a family. At Christmas time part of her present would include a new album of photographs based on her family outings from the previous year.

There was a natural progression from documenting May's life in images for her to look at in the care home to capturing her on both film and audio recordings. Marc Augé suggests our memories are shaped by what we have forgotten, that forms of "oblivion" are a necessary function in screening out our every experience. What remains is shaped by the act of being able to forget. 'One must know how to forget in order to taste the full flavor (sic) of the present, of the moment, and of expectation, but memory itself needs forgetfulness'(2004: 3). Augé uses a powerful metaphor of the garden and that memories are like the plants or flowers in it. Some need weeding out in order for others to survive. The plants that flourish have no recall of the stages of transformation from seed to flower. In order to transform there needs to be oblivion of the processes, oblivion of the action of remembrance.

Augé's work allows for the normalisation of the forgetting process and indeed highlights its importance in allowing us to process and maintain a discrete body of memory. However, forgetting becomes problematic when viewed through the lens of Alzheimer's. What of forgetting when only the distant past remains or nothing at all? How do you live in the moment and cherish the pleasures when you have no hope of recalling the events you once wanted to preserve? Technology, one could argue, has already fulfilled this function. How much of our memories of childhood would truly remain without the family album or, for the Facebook and Instagram generation, the digital archive in helping us (re)construct the story that is our life? For me, part of the recording process was an acknowledgement of the fragility of different kinds of memory and to record not just as an aid for my mother but also for myself and for future generations of the family.

## 5.2  The Autobiographical Film and Mobile Media

De Jong *et al.* in *Creative Documentary* remark on a key function of the autobiographical film being: to 'unearth hidden lives' (2012). But for what purpose one may ask? What can be learned through uncovering the highly personal accounts of our human condition? Autobiographical documentaries are suited to the technologies of the camera phone. A subjective personal approach to both documenting and documentary is foregrounded by the proximity and relationship we have with such a device. Without question these films are meaningful for the filmmaker but some (perhaps only a very few) become transposed beyond the personal realm and like past documentary traditions, seek in some way a form of transformation.

Schacter (1996) recalls how time and memory are interwoven, and that memory by definition is about the past but may often shape the future. Thus, autobiographies and who we are, who we become, is the dynamic between time and memory. In the translation of these stories we forge our identities. The autobiographical film is an attempt, in part, to capture these processes. Some memories fade through time, particularly the ordinary uneventful routines that are part of our day-to-day experience. Preserving some aspects of this provides cues for future narratives. Personal archive can contribute to a snapshot of life and in the retrieval of a memory can form coherence and a sense of belonging in a changing and unstable landscape. Making the autobiographical public in the age of the Internet can be an almost instantaneous process with very little conscious thought. However, the reflexive autobiographical film is about sharing hidden lives for a purpose, and the motives and purpose need some consideration. These may be altruistic in that sharing an event may benefit others. However, there is also something of the memorial to any autobiography in that, through time, it becomes a sign that we existed and asks that we are not forgotten.

We stop for lunch at a local pub. After we have eaten I show May some of the footage I've edited from a previous week. She can recognise herself in the scene and how she enjoys watching the New Forest ponies. There is a sense of familiarity with the image on the screen and the landscape. Whether it generates an actual memory for her is debatable. There are 'traces' of a memory enough to make the scene familiar. I start filming her with my phone and we spend time talking about the smallness of the technology. I speak about my father, how he was an avid home moviemaker and the size of the cameras he used to use. May uses my phone to look at the images and I wave to her via the lens. The camera is still running and I too am captured in the scene. We are caught up in a special moment with technology the facilitator of a mother/daughter bond. Unlike our shared memories, which will fade for both of us, this will be preserved and we can view the footage again and again. The scene is one of the rare moments I allow myself to be part of the film I'm making.

Stylistically the film was a challenge. I wanted to create something poetic and non-narrative in structure but in many ways what has evolved is perhaps a more traditional autobiographical documentary. Rain falling across the car windscreen is the opening scene and the start of the title sequence. I start with a close up as the water drops hit the glass and form rivulets which join up with others and start their journey downwards before the screen-wipers clear them away. I'm reminded of films I have seen on TV that shows the electronic nerve pulses in the brain magnified and jumping across the synapse to make connections with the target neuron. We know that memories are formed when these connections or pathways are strengthened. I remember this in the editing process and it's what shapes my decision to open with the raindrops. I don't know if an audience would get the connection but then it doesn't really matter. This is a film about lost connections and trying to forge new ones. The title sequence begins. I call the film *May Days* because these are about days out with my mother who is called May. But it is much more than that. Underlying the work is my panic as a daughter. I am losing my mother and in turn I feel lost. A Mayday is a cry for help, a distress signal sent out to anybody that can pick it up.

The next scene starts with May recalling a poem about autumn leaves. We were looking at some edited footage together and I mentioned we could see the changing seasons in some of the film clips, of time passing by. She spontaneously recalls the poem from her childhood and begins to recite it out loud. There is very little short-term memory left but photographs, music, poetry or conversation can trigger memories of May's childhood. Alzheimer's has robbed the once shy woman of inhibitions too. Her voice can be loud and she thinks nothing of reciting the poem loud enough for everyone to hear in the café. I laugh now when in the early days I would be embarrassed and try and get her to be quiet. I've done my research on the Alzheimer's website and now accept I must live in her world and accept her reality for the times we are together. May starts to tell me a story about her school days in Manchester. She recalls the road she lived on and her house opposite a police station and I know these factual glimpses into her past are correct. She talks about the times she used to go ballroom dancing and that her favourite dance was the foxtrot. I know these stories and the memories are part of her life. But then she tells me the dancing took place only yesterday and her brother, Albert, had come to collect her at the end of the dance. In the dancehall a glass chandelier hung from the ceiling and he waited for her while she danced. Apparently he is living and sleeping in a building across the road from her and they are to meet up again this evening. The memories have morphed into a strange fantasy. My uncle Albert, May's older brother died when I was a child. He tried once to teach me to dance at a wedding when I couldn't have been more than three years old. I would stand on his feet while he moved around the room. So within the fantasy there is a seed of truth in a memory of a dance that links us all.

I use repetition significantly in the edit. There are particular phrases that were repeated on our visits. 'Isn't the sea calm' when we go to the beach. 'What would the world be without birds?' as we look at the gulls hovering around the picnic tables. The New Forest ponies are: 'wonderful creatures'. May has strong links with nature. I try and incorporate in the finished edit a flavour of our encounters with the natural world against a background of sea or forest. We pull the car over to the side of the road and roll down the window. May strokes the pony who nuzzles against her through the car window. On one occasion at a picnic spot we are invaded by several donkeys who succeed in eating some of the food from our picnic. I try and get the moments on film and in the edit process I delight at May's delight in each encounter.

I used two mobile phone cameras in the making of *May Days*. I worked with an iPhone but also used an older model Blackberry. The latter camera created more of a grainy texture and had less definition than the iPhone's camera and was utilised creatively for particular scenes where the boundaries between reality and fiction were blurred; as the boundaries between reality and fiction for May became enmeshed. I was aware that what she could remember would decrease as the seasons passed. The colours in these scenes were also muted and less saturated and was used for the final scene of the film. As we look at the footage together on the laptop, the camera pulls to a close up of the screen and from there we dissolve into the scene of a group of ponies on the roadside. May is watching them from the car window. Beams from the autumn sun highlight the group against the golden background of the heathland foliage. Very slowly a chestnut pony with a white star across its muzzle breaks away from the group and walks towards the camera. May repeats the final lines of the poem about autumn leaves before stopping and saying with a laugh, that's all she can remember.

I would like to think the pony in this scene becomes more symbolic and that we are ending with something that carries the viewer beyond reality and beyond the act of remembering. May is no longer present in the final image but her voice and laugh linger. The screen shows the pony walking slowly towards the camera until his entire head fills the frame. He blinks his large eyes. Gradually the image slowly fades to white and nothingness. Unlike the physical act of remembering and forgetting, the fading to white isn't forever as with film I can start again, go back to the beginning, loop, repeat. I'm creating my own memory of my mother and the time we made a film together, my attempts capture a moment where time, in reality is running out but on film it is ever present. I am creating a shared memory which I can share with others.

The shared memory and the mediated memory are discussed in the work of Keightley and Pickering (2006) through an examination of photography and sound (specifically phonography) and how they may act as links between memory and history. Personal memories and social memories may be shaped by viewing particular images or hearing a record played on the radio. But they suggest: 'We talk of personal memories even if these are always formed in some way or another through social interaction .... So, in neither personal nor public memory is remembering simply recall. Remembering involves negotiating versions of the past, some of which may be divergent or in conflict with each other' (2006: 153).

The soundtrack to *May Days* was limited through the contracted negotiations of copyright permissions but *there is* a soundtrack in our experience and now part of our collective memories. Bing Crosby singing a selection of his greatest hits plays from an iPod as we drive around the New Forest. *Don't Fence Me In* synchs in perfect timing as we move out of the shade of the forest and on to the wide-open spaces of the heathland. These were the songs popular when my mother was a young woman. May can remember all the lyrics of the songs that formed the soundtrack of her youth. She sings along to the music, she recalls once again ballroom dancing but the story slips and slides from past to present; she was dancing 40 years ago or was it only yesterday? Between us we negotiated versions of the past we could both live with. I stopped trying to force my own version on her accepting the fallibility of my own memory too and started to accept the fluidity of time, past and present and that memories are constructed from fragments of both realities and fictions. They shape our lives and are part of who we are and what we become.

Victor Burgin in *The Remembered Film* (2004) examines how unrelated fragments of film are considered in relation to memory. Images may be recalled voluntarily but he is more interested in those provoked by external stimuli and which occur as involuntary associations. In considering memory Burgin recalls a 1977 study from the University of Provence. Interviews were recorded asking participants to describe personal memories from the time period 1930–1945. The findings were that in almost every case personal memories were entangled with descriptions of scenes from films or other such media (*ibid.*). The subjects would slip back and forth from such scenes to the lived experience until the two, in the telling of the narrative, became indistinguishable. Burgin too, recounts fragments of film which merged with his own recall where, through time, the original narratives get *loosened*. The remembered film fragments loop back to reveries, dreams or conscious reflections of lived events that may appear unconnected but have a source in the unconscious. The 'screen memory' for Burgin is brought to mind in order to conceal a repressed memory and completes or fills in the gaps (*ibid.*). He discusses, through the work of Pierre Nora, the slippage of the present into a historical past that is "gone for good". Contemporary life offers no continuity of "internal life" forged through shared everyday practices and that modern societies need to "create sites of memory" to avoid "commemorated acts being effaced from recollection" (2004: 21). Burgin advances the idea that our memories are neither reliable nor completely accurate recreations of past events but may be shaped by our media encounters. A sense of place can be constituted through our recollections, both physical encounters and at a psychic level, but the idea of place is something that can be shared with others. A war memorial in the strictest sense is a "site of memory" but an image or film scene may serve as a catalyst (2004: 22).

What I set out to do with *May Days* was a reverse process of what Burgin described in *The Remembered Film*, to create and capture fragments of everyday life and create for a moment a shared reality. I recorded them on film and in photography and so created a new short-term memory that could be played back, recalled at will via modern technology; from iPhone, to laptop, to the brain. The loops of footage mirror the loops of memory; iconic images and fleeting fragments. If you were in May's company long enough you would find the lucid conversations and topic being endlessly repeated. This was her reality and in her reality, as the film suggests, ponies are "wonderful creatures" and the sea is "always calm". The mobile phone camera became a therapeutic tool for both of us — a family tie reinforced by Burgin's ideas of creating "sites of memory" in a very literal sense. *May Days* became a film about creating identities and memories and crafting the new with the old to create a sense of belonging.

A world prescribed through Alzheimer's is both factual and fiction, it weaves real memories with elaborate fantasies. It can be pleasurable and painful. Someone once described being a family member of someone with Alzheimer's like bereavement; mourning for someone dead but still living. The creation of *May Days* gave May a sense of belonging and an understanding of who she was. For me she was very much alive and recording aspects of her life on film helped me come to terms with the processes of living with the new and the remembered. We created between us a "site of memory" that became a pleasure and a delight. I have moved from the Mayday, the cry for help in the title sequence, to May Days — days spent out with my mother — who just happened to have Alzheimer's.

## 5.3 Conclusion

In writing the chapter I wanted to illustrate how the making of a short film, using mobile phone technology can impact on the lives of ordinary people through the very personal account of the making of a film about my mother. The questions that I hope have been raised are: Why document, why tell stories, why use a mobile phone? The mobile phone as Urry and Elliott (*Mobile Lives*, 2010) have indicated is now an extension of ourselves and a technology bridging body and machine. New practices are emerging that move beyond a news gathering function on the one hand and the more prosaic uses, such as adding to our online digital photo albums or Facebook pages. There is a spectrum of creative practice in between and I would suggest, the autobiographical film — the autobiographical-documentary — has a place here. I focused too on how the autobiography could be a starting point in documenting aspects of our lives at particular moments but how at some stage the personal might move to a more public arena to create a shared experience or be of social benefit to others. By looking at how the lived experience incorporates mobile phone use, how it can be used as a tool to reinforce family ties, create memories, bond friends and add to a digital cultural and creative practice the mobile device can be celebrated as a revolutionary technology. More significantly perhaps, for the ordinary person experiencing the rapidly changing technological landscape, it can be used in a way the camera, since its invention has always been used; to underpin, or even create, a personal narrative, to *fix* a likeness and to create a memory lest it be gone for ever. So, to conclude this chapter, I will end with a quotation, which perhaps best sums up my relationship to the work I created using the mobile phone camera.

'In order to face the ongoing state of change that the internet and other digital technologies bring to filmmaking, you may wish to begin by asking not *if* new technologies change what you do but *how you might use them* to expand what you do' (Krell, 2012: 165).

# Chapter 6

# Who are Ya?: Football, Masculinity and Mobile Documentary

'Masculinity has somehow acquired a more specific, less abstract meaning than femininity...You like football? Then you also like soul music, beer, thumping people, grabbing ladies' breasts, and money...

It's easy to forget that we can pick and choose.'

(Nick Hornby, 1992: 72)

*Who are Ya?* is an early example of mobile-documentary making first screened at the Filmobile International Conference at the University of Westminster and exhibited at the London Gallery, West in 2008. It was made in collaboration with a group of young men aged 16–22. There were eight key members of the group and a much wider circle of friends who met exclusively through their shared passion of football and specifically as fans of Arsenal FC, UK. As a mobile film it forms an important reference for study as filming and editing over 4-year period encompassed new developments in camera phone capabilities and pixel size. The challenges of meeting such limitations in film quality embraced a style of edit that argued for a new aesthetic of production and reception which now can be viewed as occupying a particular historical moment in the technological evolution of mobile media. As demonstrated in the film the blurred aesthetic of the earliest filming was intercut and contrasted with the later scenes. The use of a split screen became a useful technique to overcome the limitations in picture size and format but provided a surprisingly new and unpre-dicted level of meaning on reading the finished work.

The film attempted to capture a 'visual conversation' between the football fans through the use of imagery and text messages at a time period when Arsenal FC was moving from their old ground at Highbury to the new purpose built Emirate's Stadium in north London. The chapter documents the dislocation experienced by the football fans in both coming to terms with the changes in physical space but also psychologically in adjusting to a different, sponsor driven game rooted in the dualities of economics and control. The chapter ultimately draws parallels between their exclusion from the physical space of the new stadium and the re-location through mobile and social media networks in maintaining a sense of belong-ing and fandom.

## 6.1  Context

The short film, with a running time of seven minutes, was made from collected mobile phone footage with some additional camera work shot on a small hand-held camera. The title takes its name from the chants common at UK premier league football (soccer) matches that also became a pretext for a series of research questions. Grounded by a methodology of critical creative practice (Austin and De Jong, 2008) my starting point was investigative. It was through the act of making something I aimed to explore a particular research topic. Organic in both origin and development the film evolved through stages of image gathering and editing, rather than a planned strategy from the outset. In this respect, it shares common feature of other mobile films (Filmobile Symposium, 2009). There are also similarities to the range of structures discussed by Krell (2012) in terms of 'nodal, episodic, recursive' forms and unlike traditional narrative cin-ema the nonlinear narratives of new media evolve in different ways. As such the film may be understood as a 'constellation of ideas'. I had formulated an initial research question: What does it mean to be a football fan? and through the processes of making the work further 'nodes' around belonging, shared memory and the construction of a certain kind of masculinity took shape.

The work was made in collaboration with a group of young men whose age for the project was between 16 and 22 years old. The eight key members of the group included my son and over a four-year period I collected mobile phone video footage along with text messages sent by members of the group. Much of this material was incorporated into the final work alongside my own filming.

The film opens with the football crowd outside the Emirates Stadium, the north London home to the premier league soccer club Arsenal. A split screen superimposed over the crowd follows, showing images of the former stadium in Highbury also located in north London. The screens-within-a-screen alternate with clips from football matches, still images of the fans and text messaging. My son is introduced obliquely via an image of his identity bracelet with the word 'Arsenal', rather than his name, written in Hindi on the flat part of the metal underscoring the importance of this particular team in his life.

The split screen technique served two purposes: to relay particular images that could work in juxtaposition to each other, and also to overcome the technical limitations of poor quality images produced from a particular model of mobile phone. As the technology evolved so too did the clarity of the moving images. The blurry footage was set within smaller screens either against a black background or against slow motion footage of the football crowd. Overcoming technological limitations through stylistic devices is a characteristic of some of the early mobile phone media and as such is no different from the evolution of film in general. Secondly, I used a specific technique with the screen text. It was written letter by letter, mimicking the text messages that could be sent from mobile devices. The content of these messages was directly derived from the messages the boys sent each other and reflected the colloquialisms, language and 'banter' of a football crowd. In terms of setting up a narrative, the work was edited to follow, chronologically, a football match from pre-match meeting and socialising, arrival, entering the stadium, the game, exiting and clearing up, to conflict with the police escorts and rival fans. This was embedded within a larger narrative of the relocation of Arsenal FC from their old stadium at Highbury, London to an expensive purpose built stadium sponsored by Emirates airline, less than a mile away in north London.

## 6.2 Themes

*Who are Ya?* could be described as an exploration of belonging, of what Nick Hornby has described in his 1992 novel, *Fever Pitch* as 'losing one's identity in a parallel universe'. The work is not about tracing or trying to formulate theories on particular identities but sets out to investigate how immersion in the very specifics of club football allows young men to create particular masculinities and subsume for a time other aspects of their personality. The parallel universe Hornby was referring to in the book, like the men in my film, was Arsenal football club. Hornby sets out a complex and often contradictory view of masculinity through football. Something I too, as a mother, was trying to understand in the transformation of a son into *Gooner* (current term for an Arsenal fan, used in conjunction with the term *gunner* derived from the clubs origins at the Woolwich arsenal factory in 1886 (Arsenal, 2015)). The methodology behind the work was grounded in a particular ethnography that had as a starting point the personal and the domestic (see also Spence, 1986; Spence and Holland, 1991; Kuhn, 1995).

I had watched my son around the age of seven years suddenly transform into an avid Arsenal football fan. The notion of being a fan ran parallel to the son I saw at home. As he grew older, his support for the club remained constant and shaped part of his life. I had, from the outset, a very personal agenda to discover who this new son was and what goes into creating particular masculinities. Whilst masculinity has been a rich ground for academic study (see Adams and Savran, 2002; Whitehead, 2002/2007; Reeser *et al.*, 2010), there is less written regarding the specifics of masculinities and football although some noticeable exceptions apply (Horrocks, 1995; King, 1997; Whannel in Carmichael Aitchison, 2007; Cashmore and Cleland, 2012). The key phrase in the Hornby quote referenced at the beginning of this chapter is in the idea of choice and that many men choose a very particular form of masculinity to inhabit when they attend football matches. Making a film became a starting point in my investigations rather than an end result.

## 6.2.1 *What Does it Mean to be a Football Fan?*

Football spectatorship is still very much a gendered space. Arsenal FC has a successful women's football team and there are dedicated female supporters who follow the men's team. However, what became noticeable through making the film is that the football ground and the surrounding space on match days operate as spheres of masculinity. Fans escape into these spaces and into particular kinds of performed masculinity. In some respects, one might argue, there still exists a form of masculinity which might be familiar to particular stereotypes of maleness recognisable from media descriptions and echoed in the Hornby quote. I wanted to investigate whether the men in the study conformed to certain aspects of the stereotype when gathered together specifically as football supporters. Additionally, was masculinity inhabited differently outside of such an environment where the complexities and contradictions of young men growing up in a 21st century global city were present? Hornby's autobiographical novel explores from the inside what it means to be a fan and as a writer his ability in describing and analysing such a particular sense of feeling was insightful. He challenged prevailing assumptions of the 1990s that to be an Arsenal fan was only to conform to one particular view of masculinity and that he, as a book reading male, could fully belong to the Arsenal family. My pursuit in this study is to investigate how some of the preconceived assumptions and stereotypes familiar to UK football from the 1970s, 1980s and 1990s might operate in the 21st century.

Male bonding, through a shared interest in football and following a particular team, is thought to provide a transformational or transgressive space to be free from constraints of social background, work or home life. Hornby gives an indication of the importance of this from the perspective of a fan when he speaks about "losing one's self" (1992). In such cases this might include taking on some of the stereotypical characteristics most commonly associated with the game. Some of the most consistent ideas around what it means to be a fan are often portrayed from the *outside*, that is, through media depictions rather than emanating from the fans themselves. It was important, therefore, to engage with the fans to consider such media representations and stereotypes. Some actions appeared to be consistent in match spaces such as beer drinking and banter, use of expletives and threats towards the opposing team but these negative connotations were also balanced with humour, wit and a sense of inclusion.

It is certainly true that in no other kind of sporting activity is it so acceptable to display aggression or use particular kinds of language and modes of address than at a football match. There is a tolerance of insults directed at referees, officials and opposing teams that is relatively absent in other sporting fixtures. This so-called *banter* is seen as part of the immersive and to many, enjoyable experience of attending the game. Less tolerable is the violence and racism attached to football and fandom prevalent in the 1970s and 1980s. This has been challenged with most premier league clubs running specific events to combat violence and racism in their grounds and providing education and outreach programmes. The proportion of black players at the top level of the game has significantly increased and any overt racist abuse directed at these players is no longer acceptable from clubs or fans. Whilst this is a reflection of the changing attitudes in society — to remove racism from public life — the conscious efforts at club level to educate young fans have played their part. The most prominent of these, *Kick it Out* was launched in 1993 as a joint initiative by the Commission for Racial Equality and the Professional Footballer's Association (PFA) to end racism and all forms of discrimination in football (http://www.kickitout.org, accessed July 2015).

There are of course noticeable exceptions. A case in point being the four male supporters of Chelsea FC, who were given police banning orders for allegedly racially abusing and manhandling a passenger on the Paris Metro (Davies, 2015). The images captured on CCTV and a mobile phone, were repeatedly transmitted across the globe by way of social media sites and news web channels. It served to reinforce and perpetuate particular negative images of masculinity linked to football that many had hoped were part of history, namely that to be a football fan includes racist and violent behaviour. Whilst this may be a popular misconception of 21st century fandom, media coverage can amplify the behaviour of individuals in specific incidents which in turn define aspects of 'lad culture' that are seen as concomitant with being a fan.

Homophobia is another such 'attribute' within the game and supports a type of 'football masculinity' that positions football fans as automatically 'straight' and homophobic. Today, there are still very few openly gay players at Premier League or indeed at local club level. The negative experiences of those who have spoken out have tended to reinforce the common assumptions that link the world of football to homophobia. However, a recent study in 2012 by Ellis Cashmore and Jamie Cleland presents a somewhat surprising conclusion to their research into the game suggesting it would be wrong to stigmatise the vast majority of male fans as homophobes. They suggest the suppression of gay players operate largely in the interest of the clubs and agents and their study indicates a more tolerant, liberal attitude amongst all but a small hard core of supporters. Certainly, the anonymous online interviews conducted across a range of age groups in the UK and globally — and subsequent analysis of the data — led Cashmore and Cleland to the conclusion that whilst some traditional concepts of masculinity remain in football culture, including homophobia, these are in decline and hegemonic concepts of masculinity are shifting towards a more 'inclusive masculinity' (2012).

The small sample of fans represented in *Who Are Ya?* appeared to reflect the opinions of the wider sample in the Cashmore and Cleland study. There was a metropolitan liberalism in their attitudes to sexuality and a focus on how well players performed for the team. However, I think both the Cashmore and Cleland study and my own very limited observations warrant further research. It would be interesting to compare the online responses with how fans actually behave with other supporters in the stadium space. For example, how many of the respondents had participated in homophobic songs and chants? When does banter become abuse? Banter, it is noted, acted as an umbrella term for fans for the use of language that covered everything on the scale, from witty reposts to quite serious abuse. As such it ranged from the culturally uplifting and enlightening to the most serious challenges to how language may perpetuate inequalities and would be regarded as socially harmful. Racist abuse that once might have been considered banter is now seen for the abuse it is but the same cannot be said for other forms of abuse. I would also be curious to find out if those who had engaged in homophobic chants were aware of any possible contradictions that might exist between their generally held beliefs and their stadium behaviour? That is, were they consciously performing homophobic behaviour or embodying it? I was to discover in the process of making the film that this was a starting point for formulating and researching further questions rather than getting specific answers.

Whilst attitudes to casual racism are changing and the Cashmore and Cleland study suggests homophobia is in decline, what is surprising are some of the contradictions apparent in attitudes towards women and the game. In 2015, the England women's football team reached the semi-finals of the World Cup and was very well supported yet, within the confines of football fandom, there remains a level of sexism consistent with previous eras. It is particularly interesting to consider how the football space might allow for this and also to consider how particular masculinities get maintained and performed within such space. Anthony King's 1997 study *The Lads: Masculinity and the New Consumption of Football* examined how the transformations in football, particularly the move to all-seater stadiums, affected particular constructs of masculinity. His study focused on a group of fans he termed 'the lads' whose football support and particular masculinity centred on 'drinking, singing and fighting' and who adopted a particular style of designer 'casual dress' rather than wear team colours (1997: 332). It is important to note that King, in the study, acknowledged there were many other types of fans, characterised by different masculinities, attending the matches but he was particularly interested in this group at this moment in time.

This piece of research was relevant to me as some of these characteristics, 10 years and more later, were part of my sample group namely, pre-match drinking, singing and the adoption of casual dress including the Stone Island clothing label. I also found that casual sexism was still prevalent at club level and at odds with some of the more inclusive attitudes displayed individually towards female fans.

King goes on to discuss how the solidarity between 'the lads' was forged out of a love for their team, a sense of pride, rivalry with other teams and the singing and drinking. The removal of the terraces and transformation to all-seater stadiums reduced the space where 'the lads' could congregate and immerse themselves in communal celebration. What King suggested was lost was the 'ecstatic masculine solidarity of the terraces' (1997: 336). Some have seen this kind of solidarity and masculinity equated with a working-class culture from a bygone era.

King warns against over-simplification when viewing football as a largely working-class sport and indicates how supporters have changed from one century to the next. He does, however, cite the problematics encountered in that many of the crowd have a self-belief in a working-class status, albeit, imaginary. It is those imaginary connections to particular class traditions, even outdated ones, which may inform some current practices (King, 1997). The question, therefore, is what are the conditions that allow for this shifting of attitudes from one environment to another and how do the fans come to perform particular attitudes and beliefs which are not necessarily transferred to other aspects of their lived experience including their own class backgrounds? What are the constants that make up football fandom and give them a sense of solidarity, if any, beyond the love of the team? The particular male environments that many fans choose to occupy on match days would suggest this has been significant in keeping sexism part of the football masculinity. I was corrected by a member of my study group in seeing sexism as a particular problem relating only to football. He suggested whilst sexist attitudes to women were prevalent they were not significantly exclusive and operated across many sporting environments and beyond. He argued it was more a reflection of society in general rather than specifically football. To an extent this is absolutely correct, however, I think there is something to be drawn from the particular environments of football that includes a form of sexism used to reinforce a particular type of male solidarity. Whilst it isn't exclusive to football it can thrive in that environment and be reinforced through the language, songs and banter of the stadium.

Cashmore *et al.* suggest the narrative of masculinities centred around club football and fandom offer up in the 21st century a complex account of what it means to be male and a football supporter. To return to the Cashmore and Cleland study they suggest there is evidence '… of multiple masculinities of equal cultural value in existence' (2012: 383), which offers a more pluralised, inclusive version of masculinity. Co-existing within the football space men are continuing to actively choose particular types of masculinity and subsume other aspects and beliefs as part of their experience and immersion into fandom. However, some of the more traditional hegemonic notions of masculinity and what it means to be a fan are not necessarily adhered to beyond the confines of this football space. Just as the category 'men' is not a homogenous group so too the category of 'football supporter' remains fluid. Changing social attitudes are reflected in the beliefs of many of the fans with a focus on the quality of the game rather than the race or sexuality of the player. These are strengthened further where clubs support a robust anti-discrimination policy. The performed masculinities fans 'pick and choose' for the Arsenal study group were consistent with this and embraced a multicultural dimension reflective of the group's diversity and their London city backgrounds. It would be good to think sexism within football and its supporters (and indeed all sport) will start to decline and be part of the positive aspects of the changing environment of Premier League matches. It also suggests that clubs have a certain responsibility to be active in promoting a more inclusive attitude and encourage the kind of masculinities that are not dependent on racism, sexism or homophobia that go masked under the pretext of tradition (and media assumptions) of what it is to be a fan.

## (a)  *Space and Place*

Writings on sport and masculinity have focused on the creation of camaraderie, what Roger Horrocks (1995) and others see as 'sport as a form of male hegemony'. Cashmore and Cleland (2012) and even Hornby's book *Fever Pitch* encourage a more nuanced approach to masculinity. Nevertheless, the power of football has been its ability to create a sense of community and belonging not just around a team but also a particular space. The football stadium as an enclosed space is also a place of containment and control, particularly in the newer designed stadiums like Arsenal and Wembley in north London, UK. The singing, chants, taunts and tribal rivalry, violence on occasion, continue but in a more contained form than previous points in history. Fans now have numbered seats and there are no designated standing areas nostalgically referenced as places where 'real' fans would have once congregated (King, 1997). What can be seen through the making of the film, however, is an increasing sense of dislocation, a struggle within the group identity among football supporters. This remains as relevant today as it did in the King study of 1997 and my group in 2004–2008. Who are the fans? The move from Highbury to the Emirates stadium, offered hope, excitement and a new sense of shared belonging. One of the text messages exchanged by the men in the study was an image of the new Emirates stadium with the caption "My new home". As the clip testifies, they saw the potential of the new stadium to revitalise their flagging team in the league tables but more importantly a sense of hope and possibilities to affirm their sense of belonging.

The reality, however, was different and perhaps illustrative of a new kind of displacement where corporate sponsors take control of the space and the game and less affluent supporters are priced out of the matches. There was particular hostility reserved for businesses who bought large numbers of seats for corporate hospitality but left them empty while 'real' fans couldn't obtain tickets. For those fans the surrounding streets and local public houses, in a sense, became the 'new home'. In 2012, an average adult season ticket cost between £985 and £1995 with a £15 fee charged to be placed on the waiting list. An individual match game ticket for a Premier League game ranged from £60+ to £123 depending on category (www. Arsenal.com accessed July 2012 and 2015). In 2015/2016, there remains a long waiting list for season tickets and many fans still feel home matches are too expensive to attend. Campaigns from supporters have led to some recognition at club level that this situation needs to be addressed. In 2016, it was agreed that all Barclay Premier league away match tickets would be capped over the next three years to a maximum price of £30 and Arsenal FC have reduced their price to £27 (http://www.premierleague.com/en-gb/news/news/2015-16/mar/090316-premier-league-clubs-announce-new-deal-for-away-fans.html, accessed April 2016).

*Who Are Ya?* attempted to capture this sense of dislocation and the reclaiming of street space around the stadium as their shared space to be Arsenal football supporters. Public houses, which play live broadcasts of the match on big screens, act as a gathering point for many people to enjoy the game while consuming alcohol. On match days, whilst women are not excluded from these spaces, the large majority of the clientele are male. The film concludes with a confrontation between the police and fans who are occupying the local roads around the stadium. This is really about defending one's own space, one's territory. The police tell them to "move on" but the fans remain unmoved. Why should you move on when you are in your own back yard? "Move where?" Who are these people who tell us we have to move on? These were the feelings generated and reflected through the mobile phone footage and the film demonstrates what can happen when there is a sense of camaraderie. In an act of passive resistance the football fans sit down on the pavement, in the road, in their space and in opposition to the tactics of the police who shout at them, waving batons and riot shields to move along. They sing good-naturedly "let's all sit down for the Arsenal" and the atmosphere, which might have erupted in a full-scale battle, is calmed for a time into an arena for celebration. The status quo isn't maintained and the final scenes show the police escorting many of the fans to the nearest underground station for dispersal accompanied by the sound of barking police dogs.

The footage captures a moment in time and we can hear clearly on the soundtrack the repeated voice of one of the boys filming on his mobile phone: "I'm getting it all on video, I'm getting it all on video". They are referring to their strategy in documenting their participation in a moment of heightened tension with the police. Along with the police's surveillance of their actions, the fans have the means of capturing police actions too.

There is an underlying awareness of how mobile phone footage has been used as 'evidence' in certain high profile cases where it has been alleged police acted with undue force. The report by Paul Lewis of *The Guardian* newspaper documented one such case around the death of Ian Tomlinson at the time of a G20 summit protest (Lewis, 2009). It is with the passage of time that *Who are Ya?* became, in part, a social commentary of a particular moment. The counter terrorism act of 2008: Section 76 included a clause which could make it an offence to 'publish or communicate information' with reference to the armed forces and the police (http://www.legislation.gov.uk/ ukpga/2008/28/section/76, accessed July 2015). This was taken to include aspects of photography and has blurred the boundaries of what might be possible to record in the future. Using cameras and mobile devices for the purpose of evidence gathering of police operations becomes increasingly problematic. As a consequence, the use of surveillance footage by police to combat football violence has made many fans reluctant to appear on *any* form of camera or video imaging. A climate of mistrust is easily created between law enforcers and football fans and consequently a hostile reception now awaits anyone filming extensively on mobile phones at matches. It is unlikely some of the footage captured for the film *Who are Ya?* could be so easily obtained today or that I would have such ready access to the clips shot by fans of match-day policing.

## 6.3 Form

At the heart of *Who are Ya?* is a work about mobile phone technology and a new kind of documentary making by young people; what has been called "the power of now" (Tolle, 2001) applied to the smartphone generation and their use of social media. As already indicated above the men in the study were already quite media savvy. They understood conventions of documentation and were aware how this portable and immediate access to surveillance techniques and equipment could be used for and against their own common interests. I tried to capture a sense of how football fans inside the stadium relayed aspects of the game to their friends and fans who couldn't afford a ticket. They engaged in *visual conversations* mentioned previously in the form of clips and the use of text messaging. The group of young people in this study used the visual possibilities of the phone with its camera and video function as an extension of the communication function; tools keeping them connected to friends across space and time. Images and film clips were also used in conjunction with social networking sites such as Facebook. The film attempted to embrace in its structure the sociability of image making, the exchange and the sharing and its role in constructing a sense of belonging; the role it plays in constructing personal and group memory.

Lisa Gye suggests:

> 'Photographs are often the sutures that bind the narratives of group memory.' (2007:281) She goes on to say: 'Just as text messaging has allowed people to remain in perpetual contact,...by keeping in touch through picture messaging camera phone users are able to create a shared visual space-a sense of presence created through visual intimacy.' (2007: 285)

This sharing, storing and saving images and the importance of social networking sites were all important with the age group I was working with. To be a real fan is more than just sharing the love of the game. It is about the performativity of the role itself and the acting of a particular shared kind of masculinity with others; immersion in that culture within particular physical locations. The sense of dislocation from the physical spaces of the football stadium, brought about for some through a corporate and commercially controlled game, is replaced through exchanges both visual, spoken, written in cyberspace on social networking sites. This has allowed for a continuity of feeling, the dialogue reinforces the bonds of belonging and therefore what it means to be a football fan. In this sense mobile phone culture, its uses in the specifics of creating and sharing the experience of being an Arsenal football fan and subsequently the discussions on social media sites, have facilitated this continuity. As Papacharissi has concluded: 'for most people, new media contribute to, rather than permanently dislodge, social and other routines' (2011: 308).

## (a)  *Football Chants, Music and Sound*

Premier League football still remains a gendered leisure space although the number of women attending the game is increasing. Family enclosures have also encouraged a more inclusive and diverse crowd. Yet, from the moment the match day preparation gets under-way, a particular masculine space is created and reproduced through a soundscape in and around the area surrounding the stadium. I am writing this primarily as a media practitioner and documenting my rationale for making particular choices in terms of the soundtrack for the film. Thinking through these choices provoked further questions and exposed gaps in my research. When I speak about a masculine soundscape the idea links directly to the lower registers of a male voice when compared to a female voice and through the presence of a largely male spectatorship, how this becomes amplified. The newer stadiums are designed to act as a container for this sound; the roar of the crowd can be heard outside only in its immediate vicinity. The old Highbury stadium leaked sound and every goal scored there could be heard not just in the surrounding streets but as far as Stoke Newington and Finsbury Park.

The female voice, generally of a higher frequency due the smaller size of vocal folds, is noticeable only at close range and in a stadium, not helped by the fact that the number of female supporters in the crowd is relatively small. The acoustic features of the human voice, including the timbre, allow for the extraction of socially relevant infor-mation excluding language and speech (Latinus and Belin, 2011). This includes the ability to evaluate physical characteristics including 'gender, age and size' and may, according to Latinus and Belin, be part of a 'primitive, universal and non-linguistic mode of communica-tion' (*ibid.*: 143). This idea of a primitive and non-linguistic form is of interest and worth further research. It sits alongside the more language-based chants and songs that go into creating both the spe-cific atmosphere of the football stadium and links accordingly to notions around the tribal. The colours, flag waving and drum-beating that accompany the build up to a big game have been compared to early warfare and tribalism (Percy and Taylor, 1997; Rehling, 2011).

However, it is the use of sound at the affective, neurological level to excite, exalt and transform behaviour that perhaps warrants far more investigation. The rhythmic beat of pre-match stadia music, interspersed by collective singing, generates a sense excitement. As the time to the kick-off approaches, the stewards bang shut the metal doors that separate the surrounding walkway with the entrances and exits. There is a clear divide between those inside and those outside, those who have tickets and those who do not and that is emphasised by the ritual slamming of the doors. It is also distinct not through the language uttered but by the guttural sounds that emanate from the stadium if the home team have scored a goal or if there has been a narrow miss.

I start the film with a football chant and gradually fade this out. The diegetic sound includes the noises of the stadium, the singing, the roars and the piped pop music. The non-diegetic sound includes the music track: *Somewhere* by Leonard Bernstein and Stephen Sondheim, (1956) for which I received copyright clearance for a limited and conditional period. It is played on a solo saxophone, first slowly and rather melancholic then as a jazz version. For those who might recognise the tune the opening bars: "There's a place for us…" was intended to juxtapose images capturing the enthusiasm for the move to Arsenal's new home at the Emirates stadium, which subsequently turned to disillusionment.

## (b) *Mobile Documentary*

One of questions for myself and other practitioners and academics who took part in the *Filmobile* and *Mobilefest* Conference (2008) (see also Schleser, 2013) and *Documentary Now* Conference, London (2009) was around the subject of documentary practice and mobile filming and whether new cultural patterns and aesthetics were emerging. Some current debates have polarised opinion in terms of mobile phone technology and the aesthetics and ethics of filmmaking. One position tended to valorise the blurry images and shaky hand-held techniques; the technical limitations celebrated as aesthetic strengths.

Particular codes of filming and editing emerged to counteract the technical weaknesses but as these improved so too did the creative practice and form. The juxtaposition of blurry images with other footage, use of multiple framing and 'frame within a frame', etc. are some of the editing techniques used to overcome such limitations in the technology and were strategies I employed in *Who are Ya?*

In newsgathering the blurry nature of mobile phone images taken by amateurs are often validated and legitimised as more newsworthy than professional images. They are increasingly used on the web and in TV news programmes in the first round of images to support a breaking story. This stems in part from the perceived immediacy of the medium in the form of eyewitness accounts. Examples can be drawn from 9/11, New York, the July 7, terrorist attacks, 2005 in the UK and tsunami footage from Japan in 2011. Audiences have become familiar with the wobbly camera and have quickly learnt to read the text as 'authentic', eyewitness accounts of an event happening in the 'real' world. So, in this respect, one might argue these codes of construction have been validated through their use as authentic and immediate. They have acquired currency through their legitimisation on TV news programmes and certain kind of documentary practice above more polished and refined filming and editing techniques using professional equipment. Drama and feature films have utilised these particular methods with *The Blair Witch Project* (1999) and *Cloverfield* (2008) to create a sense of the 'real', 'immediacy' and 'authenticity' as part of their narrative technique.

An alternative position is taken up by Steve Hawley (2008) that compares the aesthetics of the camera phone to early examples of television. In this sense the use of close-ups, repetition, blurry footage, biography, banality, is not a new aesthetic, merely a phase in the development of mobile media. As the technology improves, he suggests, there will be little to separate the aesthetics of mobile films and the more conventionally shot work. We are at an interesting stage in mobile media aesthetics where both views have some currency. The latest smartphones come with advanced camera technology that produces professional looking images so the days of blurry footage may be coming to an end. However, what really sets the mobile phone camera apart is its size, accessibility and connectivity to a range of social media that has changed the dynamics certainly at the level of distribution.

In this respect, Thomas Meyer (2008) who speaks about the mobile phone being "an enabler of social interaction in everyday life" gives an indication of the kind of uses it is best put to when considering creative media. As a tool, it can bear witness to things and amplifies the opportunities for picture making. The potential for how camera phones are used is vast. It lends itself particularly to the hybrid and more personal documentary genres.

## 6.4 Conclusion

In this chapter, I set out to discuss the methodology and ideas behind the making of the mobile-documentary *Who are Ya?* (Fig. 6.1). In doing so, I have attempted to explore what it means to be a fan of a north London Premier League club through the images, texting and discussions among a group of supporters that included my son. Through this I attempted to understand the complexities and the nuances involved in construction, performing and maintaining masculinities, both hegemonic and inclusive, related to a specific geographical location, space and time. This stemmed from a personal rationale rooted in domestic and family ethnography. However, in my exploration I came to conclude there are many types of family. To be a football fan is to be part of another kind of family and the sense of belonging this generates may be different but the ties are just as strong. There is a sense of feeling and shared collective memories that bind supporters together. The use of social media, in this respect, has helped to maintain bonds across cyberspace and geographical distance. The sense of belonging can transcend the divides of class, race and gender. Certain traditions may be transferred from one generation to the next and can be witnessed in the continuity of the chants, songs and banter but as cultural meanings get reinforced so, conversely, they can weaken. Whilst 'hard-core' views are undoubtedly held, for many there is greater tolerance to differences expressed at an individual level with evidence to suggest that racism, homophobia and extreme violence appear to be in decline. This is supported by home office statistics that indicate a reduction in the number of football related banning orders and arrests made in England and Wales. Banning orders have decreased by 7% and arrests are: 'The lowest on record' for the 2013–2014 season and continue part of a downward trend (https://www.gov.uk/government/uploads/system/uploads/attachment_data/file/352864/FootballRelatedArrestsBanningOrderSeason201314.pdf, accessed May 2015).

The biggest 'enemy' as identified by the fans is capital and the corporate development of the game. The quote from Goldblatt best summarises what football fandom means at the lived, everyday level and how cultural meaning becomes embedded within particular practices.

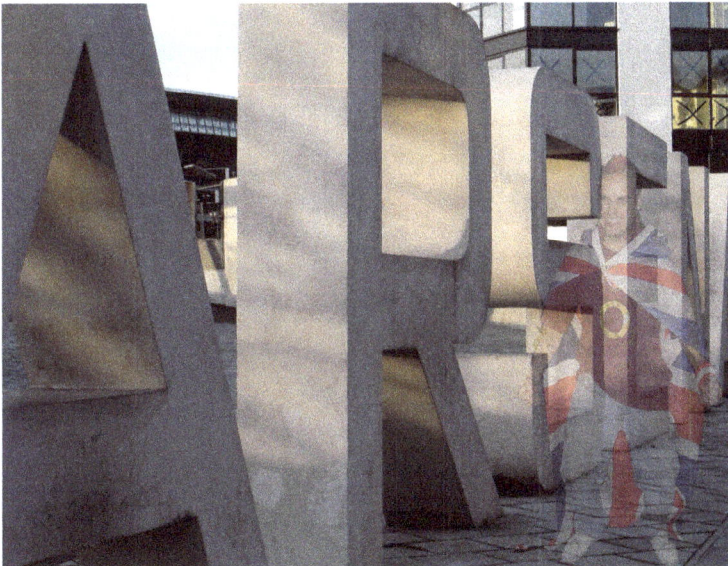

Fig. 6.1   Publicity image from the mobile-documentary *Who are Ya?* Sylvie Prasad (2008).

'People who love Arsenal, when you buy into Arsenal and you start supporting Arsenal, you're not just supporting the Arsenal of now, you're supporting and embracing a whole series of cultural meanings,

stories of the past which cumulatively constitute what Arsenal is.'
(Goldblatt, 2014)

I also wanted to examine the aesthetics of the mobile-documentary through the techniques I used to overcome the limitations of the early camera phones. This was set against the debates of the time which polarised discussions as either a celebration of these limitations and the new codes and conventions that emerged as a result or viewed as merely an annoying phase in the technological development of mobile-film making. *Who are Ya?*, in a sense, maps the developing technology of the camera phone from the first blurry footage to the broadcast quality we now have. It forced me to think creatively on how the work might be exhibited which, in turn, led to the technique of the visual conversation, split screen and text messaging. Such constraints, rather than a hindrance, became catalysts that allowed the creative process to flourish. The final work attempts to capture the spirit of how the subjects in the film actually used their phones in an act of sharing information, which bound them together as football fans and captured a particular moment in time.

# Chapter 7

# The *Selfie* and Media Advocacy

'Definition of selfie in English:
noun (plural selfies)
*informal*
A photograph that one has taken of oneself, typically one taken with
a smartphone or webcam and shared via social media.'
(http://www.oxforddictionaries.com/definition/english/selfie).

> '…our loathsome society rushed, like Narcissus, to contemplate its
> trivial image on the metallic plate. A form of lunacy, an extraordi-
> nary fanaticism, took hold of these new sun worshippers.' (Charles
> Baudelaire 1821–1867, *The Modern Public and Photography* in
> Trachtenberg, 1980: 86)

On November 19, 2013, Oxford Dictionaries announced the
*Selfie* as their international Word of the Year. The editorial director at
the time, Judy Pearsall, explained the phenomenal upward trend in
the use of *selfie* across the year resulting in its selection. The decision
to include the word in the dictionary, particularly as word-of-the-year,
was not without controversy. Reports in the press and Internet media
tended to fall into opposing camps, polarising the debates regarding
the merits of the selfie that continue to this day.

The starting point of this chapter is that the current media preoccupation with the selfie is an indication of a more narcissistic, individualistic society shaped increasingly by our relationship to a celebrity-driven

culture. It suggest, however, that through a study of what it means to document the self via social media, a more complex narrative emerges and ultimately links us back to questions about what it is to be human. The chapter continues by discussing the use of mobile phone photography and the selfie and how some academics have responded to media critiques by calling for a more nuanced reading of the phenomena. The chapter concludes with an approach to creating student projects that critically engage with the selfie as a mode and means of connectivity and media advocacy.

## 7.1 Debates

Two images have gained iconic status and focus of both media and academic discourse on the subject of the selfie. In 2013, professional photographer, Roberto Schmid, captured at the memorial service for Nelson Mandela, US president Barack Obama, Danish Prime Minister, Helle Thorning-Schmidt and UK Prime Minister, David Cameron posing for a selfie. The photograph was featured on the front page of news stories around the world, went viral on Twitter and other social media and brought in its wake numerous comments about the impropriety of taking such a picture at a memorial service. The photograph also depicted to the right of the frame what has been described as an 'unhappy' Michelle Obama and fuelled media speculation about her apparent disapproval of her husband's behaviour. Questions were raised in respect of all three politicians as to their suitability for leadership (Soltis, 2013; Anthony, 2013; Beattie, 2013). The second iconic image and slightly less contentious selfie, was taken at the 2014 Academy Awards ceremony in Los Angeles and posted on social media site, Twitter by Ellen DeGeneres, the show's host (Addley, 2014). It has become, to date, one of the most retweeted selfies of all time. We might consider what it is about the selfie as a visual mode of representation that appeals to such a wide cross-section of the world and ask ourselves who hasn't taken a selfie at some time or other? No other form of photographic practice has received such vehement criticism or evoked such extremes of opinion.

The media continues to report on the selfie and in April 2015 Jonathan Jones' article for *The Guardian* newspaper declared: 'RIP the selfie: When Prince Harry calls time on a craze, you know it's well and truly dead' (Jones, 2015a). It seems even members of the British Royal family feel obliged to voice an opinion on the subject. Further news coverage followed with museums and art galleries around the world joining the debate on the value or otherwise of the selfie. Some of these institutions, including The National Gallery in the UK, banned the use of selfie sticks arguing that it was a question of safety and that people should be looking at and appreciating the works of art on display rather than taking pictures of themselves in front of the exhibits. This approach was counter-balanced by the opening of *Art in Island* in the Philippines. This was the first gallery specifically designed for taking selfies with the art on display and described itself as the first 'Selfie Museum' (Wyatt, 2015).

Underlying a good many of these public comments are judgements about culture, society, the function of art, the state of the nation, technology and moral panics about young people in general. Many critics position selfie culture as symptomatic of a narcissistic society in which people, engrossed in a technological bubble, are losing the ability to experience and relate to anything in the 'real' world. Historian Simon Schama best summed up this perspective in 2015 with his attack on the selfie outside the National Portrait Gallery, London, at the launch of his exhibition, *The Face of Britain*. Mark Brown, arts correspondent for *The Guardian* reported Schama had declared: "society would be a better place if people, perhaps on their daily commute actually looked at the faces of strangers" (2015). Whilst the majority of media coverage had a distinctly negative opinion of the selfie and its significance, a number of published articles offered a more positive evaluation.

Bridie Jabour's online report (2015): *Hate selfie sticks, or just young women having fun*, took a distinctly feminist reading of the selfie phenomena. She attacked the numerous critiques, which, she suggested, were written from positions of power. She comments: 'What exactly is it about selfies that raises the ire of people who have built themselves lofty pedestals from which to dictate to people what is correct and acceptable behaviour?' (*ibid.*). She goes on to state that young women tend, most often, to be the target of "selfie scorn" and selfies may be a way for a woman to take back control of their image in an age where real empowerment for them is limited.

There may be some currency in this position. *SelfieCity* investigated visual data across the cities of Bangkok, Berlin, Moscow, New York and Sao Paulo covering five continents and found, across a range of 32,000 images, that it is both a young person's activity (the median age across the sample was 23.7 years) and more women than men post selfies on social media websites (http://selfiecity.net/, accessed July 2015). The young and young women in particular, may well be easy targets for media criticism. However, the Jabour article was one of the first popular news stories that hinted at an underlying prejudice at work in the derision of the selfie that taps into wider questions about the value of particular kinds of cultural expression. It also suggested that taking and distributing selfies online may not have the same rationale for everyone and we might think about how the selfie becomes a gendered practice. We will look at some of these ideas further in the chapter.

## 7.2  The Social Selfie

What is different about the selfie from other forms of self-portraiture is the immediacy of the technology; to make instant photographs in any place or setting and share those images on a global scale via social media. This is significant and sometimes overlooked when making historical comparisons with self-portraiture. Many of the selfies are used on sites such as Instagram, Twitter, Facebook and Tumblr. The ability to share selfies instantaneously is appealing but can have both positive and negative consequences. It is often the latter that becomes a focus for media attention. *Selfies at Serious Places,* a link on Tumblr. com, documents what is seen as a particular troubling aspect of the genre. Selfies posted from holocaust memorials and concentration camp sites feature alongside more banal images. There is also a Tumblr link to *Selfies Taken at Funerals.* The backlash against such imagery has certainly triggered further debate leading to some contributors to the website publicly documenting their apologies. These examples of selfies are usually invoked in a climate of media and moral panic against young people with rhetoric that positions them as self-absorbed, celebrity-driven and lacking any civic or social responsibilities.

Jason Feifer (2013) who compiled the Tumblr site commented:

'There's a lot of youthful stupidity on display here, but every prior generation would have embarrassed themselves publicly, were they equipped with the technology to do so. This Tumblr captures people in moments they haven't fully thought through, but that doesn't mean they're incapable of thinking further.' (http://selfiesatsriousplaces.tumblr.com/, accessed July 2015).

A popular online commentary written by Jerry Saltz that examines the history of the selfie describes one of the images with a tag line: "*Selfie taken from the gas chambers at Auschiwitz*" and a second post captioned "*The photos are of me at Treblinka…*" These kinds of selfies provoke some of the most severe criticism. However, Saltz attempts to position this category of selfie alongside certain kinds of art practice and highlights the need to investigate just how different age groups engage with social media. He comments:

'We can't merely dismiss these as violations of sanctified spaces or lapses of judgement. Atget photographed crime scenes. War correspondents catch images of people being blow to bits… Many [images] are in bad taste, and some indulge in shock value for shock values sake, but they are, nevertheless, reactions to death, fear, confusion, terror, annihilation.' (2014: 3).

This, indeed, is a different perspective to the usual media coverage on selfies. Here is an attempt to try and investigate what lies beyond the surface of the images, the actions behind them and their contexts. This is perhaps the most useful of approaches as it focuses on trying to understand the complex relationships humans have to themselves, their environment, culture and society as opposed to the 'blame and shame' approach adopted by some media sources. What we also need to consider is how digital media may be changing the social landscape and if this apparent dislocation between the past and the current lived experience young people inhabit is a reality. Let us consider the response a young man gave to his critics after he had posted a selfie from Chernobyl on June 5, 2013:

'... I was fascinated by what I had heard about the event — as I'm only 20 I wasn't around when the disaster unfolded, so keen to learn more I took it upon myself to go to Ukraine and visit the site…That expression on my face is meant to be shock, not some vacuous, feeble attempt at narcissistic irony. The shock of seeing the scene of the disaster I'd heard so much about. The shock of seeing how close the reactor is knowing that 27 years ago standing in the same spot would have killed you in mere minutes….All these feelings culminate inside you and you're left numb and confused trying to take everything in. It was probably one of the most surreal experiences for me during my time on Earth, without question… Finally, I think some people have been quick to demonise myself and my picture. As I've already said, you don't go to Chernobyl for a party, and I'm quite proud that a photo of me taking in some culture, documenting my travels and learning about the world's worst nuclear disaster has gone viral.' (Fletcher, 2013: available at http://selfiesatseriousplaces.tumblr.com/, accessed July 2015).

It can be argued that the apologies, rationale and responses from young people who posted on sites such as Tumblr came as a result of the criticism directed at them. However, as the Chernobyl example shows, what cannot be argued is that they are disengaged. Furthermore, they have the ability to think about the world and their place within it as is evident from their responses to the criticism directed at them. As others have pointed out, part of being a young person is to adapt and change as one gains experience.

'Instantaneous' and 'public' are two of the attributes that smartphone technology is credited with. We can be part of a global networked community where our actions can be speedily disseminated. Equally the consequences of those actions can have greater resonance when visible to a global community. 'Trial by Twitter' has been at the forefront of a number of media stories, and not all featuring the younger generation (McDonald, 2012; Johnston, 2015). There have been large-scale, collective responses to events, actions or individuals on social media that have, generally speaking, been opinion driven with very little regard to evidence. The effects of such social media 'trials' can have serious repercussions particularly as the posted comments are almost impossible to erase, even if they prove to be incorrect.

Transgressions are considered inevitable on the journey from youth to maturity. The ability to forget, or at least forgive, these transgressions is an inherent part of the process. The digital world has changed this and now at the heart of social media is a global visibility. Transgressive behaviour, rather than being the concern of the family or local community, becomes part of a much wider, global online debate often leading to personal attacks from people outside those circles. Online data remains retrievable *ad infinitum* so the prospect of living a life, and inevitably making mistakes, in relative privacy becomes impossible. So the question worth asking is: How might human beings adapt and evolve in such a world?

As I write this chapter, the following news story has been broadcast by BBC News, '*Sexting boy's naked selfie recorded as a crime by police*' (http://www.bbc.co.uk/news/uk-34136388, accessed September 2015).

The mother of the 14-year-old boy at the centre of this story described her son's action of sending a naked selfie to a girl in his class at school as naïve and thought his public humiliation punishment enough. However, the boy, branded a criminal by the police, could have his details stored by them for the next 10 years and made accessible to any future employer. The tensions in negotiating a cyber culture are apparent from these examples. We are living through a speed of change and technological development that has been likened to the rapid transformation experienced in modernity. Looking back we can see how those major shifts impacted on every aspects of life, economically, socially, culturally and in our relationship to time itself and how we were constituted as human beings (Berman, 2010). The full implication of our digital transformation remains a matter of speculation.

How the Internet is used and for what purposes, along with the lack of transparency regarding data collection, has raised a number of issues regarding public safety online. In turn, a number of initiatives have been developed. The *5Rights* campaign is a civil society initiative aimed at empowering children (under 18s) by 'delivering a universal framework of digital rights'. It has backing from the UK Government as well as numerous organisations and charities. One of the rights is the *Right to Remove,* which will allow children and young people to edit or delete any online content. The website carries the following explanation:

> 'Errors of judgment, unhappy experiences and attitudes that were the product of immaturity are saved on the internet long after they have faded from the memory of friends and family. This can make it extremely difficult for young people to get away from their past experience and move on. Online, their past coexists with their present, leaving an outdated, un-contextualised public record.' (http://5rightsframework.com/, accessed June 2016).

Safeguarding our freedoms and rights, particularly where young people are concerned, are seen as important criteria to enable societies to develop codes of conduct for online actives. However, this also highlights some of the problems in trying to protect the vulnerable but keep an open network. It also raises important issues about the nature of power relations and who governs and regulates cyber culture and for what purpose. The following section draws attention to studies which suggest young people may relate to cyber culture in a different way to an older generation. It also draws attention to how inequalities in society, particularly in terms of age and gender, may operate and perpetuate across the world wide web.

## (a)  *Selfies, the Young, Gender and Control*

Whilst generations have always exchanged sexually explicit letters, images, messages and other material, such activity becomes contentious in debates about children and young people. They are both commodities within the pornographic industries and at the centre of many policies and laws established to protect them from exploitation. It is an oft-quoted fact that over a third of all internet use is in the consumption of pornography although this figure is contested (Ward, 2013). However, what is not in doubt is that that the production and consumption of pornography is big business and there is a thriving online culture that exploits children.

*Sexting*, sending images and/or text messages with explicit sexual content via a mobile phone, can be seen as part of a continuing set of practices to exchange, buy and sell explicit material. Where it differs is in time (instantaneous), scale (global) and the digital footprint that is left behind. As we have already noted, it is very difficult to remove material from the Internet once it is in circulation. Moreover, there is very little control over how the images and messages are used and re-appropriated once on the Internet. Whilst it is illegal in many countries to send nude or explicit material of anyone under the age of 18, *sexting* or sending explicit selfies often involves teenagers. Some media stories have reflected concern in the increase of sexting among the young and how it has, in some cases, been used as part of a cyber bullying culture. A growing number of websites providing information on how to 'stay safe' on the web are being established and while no one would argue about the need to keep children and young people safe from harm and exploitation, the discussions around *sexting* by teenagers have not been so straightforward. Elizabeth Englander, writing in *The Washington Post*, suggests that *sexting* is 'becoming a normal part of a teens' sexual development' (Englander, 2014). As such, there is a difference between curiosity and coercion and for Englander, the focus needs to be in educating teenagers to understand this aspect rather than blanket condemnation.

We now turn to academic studies on the selfies that offer a contrast to the more journalistic debates we have covered so far. Kath Albury in: *Selfies, Sexts, and Sneaky Hats: Young People's Understandings of Gendered Practices of Self-Representation* suggested, through her study of a group of 16–17 year old Australian subjects, a gap between the teenagers interpretations of their online digital practices and those of adults. The study was primarily focused around a discussion of *sexting* and the production and sharing of naked and semi-naked selfies. Albury's study made clear the group were conscious of the difference between private selfies, that is forms of sexual self-representation as a mode of self-refection, in contrast to public selfies: a "self that communicated to others" (2015: 3). Of the latter, the conclusion was that these kinds of images, in particular the more "jokey" variety, were more likely to be misunderstood by adults who, not surprisingly, wouldn't get the joke. More significantly, Albury, in her final analysis points to double standards in respect of responses to male and female bodies. She argues: "While young men's naked or semi naked selfies can be dismissed by both adults and young people as merely funny or stupid, young women's participation in selfie culture bears an extra burden of representation" (2015: 9). In the study Albury notes that sexual self-representation became highly problematic for young women as they were more likely to undergo public censure and shaming than young men.

This is taken further in investigations by Anne Burns who argues: "The gendering of selfies and the prevalent assumption that they act as proof of narcissism limit how women's personal photography is socially valued" (2015: 8). Burns takes a Foucaultian approach to the reading of the selfie phenomena and how they are discursively constructed as a gendered practice. This in turn allows them to be further devalued by associations with the feminine in terms of "triviality and vanity". She identifies two particular selfies that emphasise this notion. One depicts the image of astronaut Neil Armstrong with the caption: 'Went to the moon, took 5 photos", the second is of a young woman with the text: "Went to the bathroom, took 37 photos". The juxtaposition of the two images, Burns suggests, counterpoints Armstrong's achievements and restraint with a woman's "implied obsession with triviality".

Whilst, to an extent, all selfie takers are subjected to a categorisation that judges their images as trivia, Burns suggests this is a: "shorthand to chastise those whose photographic self-depiction is perceived as self-absorbed or crass" (2015: 5). This has further implications for women as it becomes, in Burn's view, part of a set of regulatory practices aimed at controlling the female body. This perspective gives us some key points to think about as it focuses on the perpetuation of particular stereotypes of women, normalising the discipline and punishment of those who don't fit the 'norm'. We might also consider from these conclusions how particular hierarchies are maintained in an online public sphere.

## 7.3  Capturing a Likeness

There are genuine fears about how technology changes the world that underlie many of the debates fuelled by a media hype that shows no signs of abating. In the readiness to criticise certain section of society in relation to the types of cultural practices that are adopted, this is nothing new. These debates may be viewed as part of a continuum and therefore it is interesting to trace some of the current thinking in relation to their historical foundations. Many scholars locate the roots of the selfie in the practice of both vernacular photography and self-portraiture (Rawlings, 2013). They cite the first Daguerreotypes and what is thought to be one of the first self-portraits taken by Robert Cornelius, a former metal worker of Philadelphia. Other self-portraits, (see Joseph Byron, New York 1909, film director Stanley Kubrick, 1949), strongly suggest the practice is as old as photography itself (Stubbs, 2014). Prior to the emergence of photography, portrait painting was popular amongst those who could afford it and an extension of this included many artists capturing, in paint, a likeness of themselves. Silhouette cutting, another precursor of photography, became fashionable among the expanding middle classes of the late 18th century for capturing a likeness (Wells, 1997/2015). Moreover, the earliest records indicate that there seems to be a desire to see one's self, to look at one's self. Photographic theory has turned to the work of Sigmund Freud and Jacques Lacan for help in understanding the relationship to looking, seeing and being seen. Freud documented the pleasure of looking (scopophilia) as an autoerotic function in a child's development. The ideas were expanded into the 'mirror stage' by

French psychoanalyst Lacan and the *mis*recognition and alienation of the self (Bate, 2009).

> 'Thus human identity (social, sexual, political) is always a precarious structure, precisely an *identification* (process) that is subject to 'others'. In this respect we might see that a central gratification of portraiture is precisely an address to the imaginary question: am I like this person or not?' (2009: 82)

Bate outlines how some photographic theory has turned to psychoanalysis to explain that identification and visual pleasure are central to the spectatorship of portraiture. By extension we can apply these to self-portraiture as part of the desires to know one's self and to have that self-seen and verified by others. Photography, like painting, is also about a constructed self and some of the earliest portraiture was connected to a 'valorisation' of the self. By that I mean it had a social function and was purposely constructed for others. We might pause to draw comparisons here with the selfie and how technology has enabled us to 'play' with different versions of ourselves, which by way of the Internet, we construct for others to see and validate.

Early portraiture involved visiting a studio and having a photograph taken by a professional photographer. Travelling mobile studios took photography to outlying towns and villages and were popular in helping to expand the practice from being affordable only for the wealthy to those with limited funds. Props and smart clothing were made available and a variety of backdrops could be chosen to enhance the image and give an air of social standing irrespective of personal background or circumstances. The photographs were very stylised and control of the image lay with the photographer who decided how subjects should pose. In February, 1900 Kodak introduced its first hand-held Brownie camera and launched what quickly became a mass-market technology. Viewed as a 'democratising technology', the Brownie was simple to operate and priced at just $1 within the reaches of ordinary people. (http://www.kodak.com/ek/us/en/corp/aboutus/heritage/milestones/default.htm, accessed April 2016).

For the first time people had control over what to photograph and how they wanted to be represented. Whilst the 'idealised' image, mimicking the photographic codes of the professionally taken photograph and evidenced in family albums continued to present 'edited' and more formal versions of family events, it gradually gave way to a more relaxed, informal approach. The snapshot, considered to be the most spontaneous, casual and unconscious of all genres of photography, was linked to both mass market and domestic genres. This casual aspect of the snap could also, according to Val Williams (1991), be seen in the light of specific codes of construction. She writes how ideas about the family emerge through the snapshot far more than other photographic genres where they are located and placed within the present. There is a process at work and identifiable, which selects from the 'hum drum' and the everyday, to give a cogent view of domesticity. It is through such a selection process that we can start to analyse what was considered of value and had meaning and how this may be different for different groups in society. How we see the world, how we see others and how we see ourselves forms some of the key discussions in theories of representation but they can also be examined in terms of how particular practices can be valued or not.

The Kodak advertising campaign "You press the button, we do the rest" gave the impression no skill was required. It certainly separated the taking of the photograph from the more difficult chemical processes of developing and printing (Wells, 2015). This added to the growing divide between the masses; vernacular photography and the more 'artistic' pursuits of the keen amateur who still remained in control of both image capture and printing.

The ubiquity of the selfie has fallen into many of the above categories. It is seen as requiring little skill or thought. It has also become part of a set of popular cultural practices and embraces the vernacular that, for some, are not in keeping with other forms of photographic art or, indeed, art in general. Furthermore, it is associated with a young and largely female practice, which serves to amplify gender inequalities and prejudices. On the other hand, as part of a global phenomenon, it cannot be so readily dismissed. Within the academic community there is call for a more nuanced approach to how the selfie is considered. Nancy Bayn and Theresa Senft, founder of *The Selfie Researchers Network,* comment:

> 'When people pose for political selfies, joke selfies, sports-related selfies, fan-related selfies, illness-related selfies, soldier selfies, crime-related selfies, selfies at funerals, or selfies at places like museums, we need more accurate language than that afforded by 19th-century psychoanalysis to speak about what people believe themselves to be doing, and what response they are hoping to elicit.' (Senft and Baym, 2015)

They suggest that focusing on references to the historical, in particular the more Freudian based comparisons, may not be the most productive in relating to current practices. The following section attempts to engage with some more forward-looking ideas and analyses that take account of the digital environment and the specifics of the new technology.

## 7.4  What is a Selfie?

The chapter began with the Oxford English Dictionary's definition of a selfie and this section examines further what this actually means in terms of practice. The first selfies taken on mobile phone cameras were close-ups of the face and tended to be somewhat blurry, a fact that emphasised, more than anything else, the limitations of the technology. They were taken at arm's length and particular codes of construction were adopted from the more conventional photographic portraiture; tilts of the head, subjects looking to the left or right or the classic 'mug shot'. Props including hats and sunglasses started to be used and the genre evolved into specific sub categories: *selfies with hats, selfies with hand gestures, travel selfies,* etc. As the camera technology, including the automatic focus improved, more variety developed. Some of the most common selfies include aspects of the background, so gives the viewer a sense of location or place. Indeed, this aspect of the selfie is seen as part of its appeal for the online audience. We get glimpses into the bedrooms, sitting rooms, holiday destinations, towns and cities of the selfie-taker. We are able to locate the subject both in the familiar of the domestic setting or the unfamiliar of an exotic location. There is a voyeuristic pleasure in trying to piece together a life in a game of compare and contrast across distance and time.

Frosh (2015) argues the selfie is a form of "relational positioning". The individual bodies of the viewer and the viewed are linked through the network in a mutual but shifting connectivity. He terms the selfie a "gestural image" in that within the photograph there is always an imprint of the actions of the photographer. This can be as obvious as seeing part of the arm (or selfie stick) within the frame. To take a selfie always implies an action or movement of the body, which doesn't come naturally; it is part of a connected performance (2015: 3). In this way, a selfie is presented as a more intimate image than other forms of self-portraiture. As viewers we see the construction and performance of the self, simultaneously photographing themselves.

## (a)  *The Celebrity Selfie*

Interestingly, the celebrity selfie may not offer the same appeal as the more informal images taken by non-celebrities. In the case of the latter there is a certain pleasure to be had at the lack of photographic skills with, for example, aspects of domestic life made inadvertently visible in the peripheries of the shots. Celebrities use the taking and sharing of selfies as a way of keeping connected with their fans and followers but very few really allow any of us into their personal space and so reveal anything we might not already know about the person.

Kim Kardashian, celebrity of US reality television *Keeping Up with the Kardashians* and prolific selfie poster on social media, is a case in point. Singled out for both media and academic scrutiny for her now famous *belfie* (selfie plus bottom) that had Kardashian posing at home in a white swimsuit revealing her body but shows little or nothing of the space in which it was taken. What we see is a photograph that follows all the codes and conventions of a celebrity body image intended 'to extend and maintain the celebrity of its subject' (Mirzoeff, 2015: 66). Celebrity selfies in this respect are best considered as a different category to the 'ordinary' selfie, fulfilling different functions for both the viewer and the viewed. Mirzoeff equates these kinds of celebrity selfies as an extension of the 'film still' or advertising image. They pretend to be the exclusive work of the celebrity but are in fact part of the continuum of images that form publicity material. As such they are often 'managed' by a set of 'cultural intermediaries'; publicists, managers and agents who control the media image of their clients. Even when the celebrities have full control over their Instagram output the images released to the public are highly manipulated through the use of pre-sets or filters. Many of the Kardashian selfies were collated and published as a book in May 2015, further emphasising how the celebrity body is the commodity subject to carefully edited and regulated control as part of an income stream for the brand Kardashian. It is therefore important to examine the celebrity selfie not only from an economic perspective but also from the techniques employed in the construction of the images; how they use conventional, highly polished production values in terms of lighting, composition and poses common to advertising. It is also worth asking just how much of the self is actually part of the celebrity selfie and consider how the codes may, in turn, influence the construction of non-celebrity selfies and in particular those taken by ordinary women.

## (b) *The Travel Selfie*

The invention of the selfie stick extended the distance between the camera and subject. This allowed for greater flexibility in positioning the camera, enabling it to capture more of the background and lent itself to the group shot. The travel selfie, an extension of tourist photography, positions the subject in a particular location in which backgrounds are an important referent. It is not uncommon to see people carefully posing for a group shot at major tourist landmarks. The selfie taken alongside popular, iconic locations such as the Eiffel tower in Paris or Big Ben in London, are a particular sub-genre and these kinds of selfie are replacing the picture postcard. They form a record of the 'sights seen' but validate the experience by the insertion of the holidaymaker into the landscape. Aaron Hess suggests: 'Selfies invite consideration of the composition of the self in space as digitally and visually rendered. In other words, selfies visualise the user as emplaced within the physical surroundings and as digitally embedded into social networks' (2015: 8). This dual function of the selfie sets it apart from other tourist or travel self-portraits taken on cameras without the smartphone capabilities to network.

This kind of tourist self-portrait is often criticised as a lack of engagement with the physical, geographical space and place and that the view is always tempered through the lens of a camera. This is not just specific to mobile photography but was raised in 1977 by Susan Sontag in her seminal book *On Photography.* One of her essays debates the rising concerns on whether an image-world was replacing a 'real' world, that photographs provide a second-hand information without undergoing the experience. Sontag makes very clear the complexity in trying to define a particular, single notion of reality but goes on to suggest that photographs are a way of 'imprisoning reality' and of making it 'stand still' (1977: 163). It is a form of participation in the world but at a comfortable, sanitised distance.

These arguments have taken a more current turn as we saw in the latest critiques around mobile phone photography and in particular the selfie. There is certainly validity in wanting to 'capture the moment' before it is lost forever. There are echoes to a previous era here and how photography emerged as one of a set of practices best suited to documenting the rapid changes of modernity. They formed part of a permanent record of a people and a landscape transformed and the shifts in culture in the move from the agrarian to the industrial. In the postmodern age of cyber culture, information overload and fears of an aging population at risk of dementia, taking a selfie consolidates the idea that *this is me* and *I was here.* What is different is that selfie takers reinforce this sense of existence by the instantaneous sharing via social media, which moves beyond the image and into a networked community. Selfies are rarely ever made just for personal consumption. They begin a conversation with others so rather than viewing the selfie, the image, as an escape from the material world they are extending this experience into a parallel cyber world networked not only to friends and family but also to strangers. The worlds co-exist rather than one replacing another. As Tifentale (2014) comments, the act of sharing a selfie sends signals about belonging or wishing to belong to a community.

Hess, like some of the previous scholars featured in this chapter, calls for a deeper reading of the selfie as it may point to a more insightful account of the relationship between technology, networks and ourselves as humans. In his account, he takes a more Deleuzian approach when he speaks about 'selfie assemblages'. By this he is referring to a dynamic nexus between: 'intimate self, public spaces, locative technology, and digital social networks' (2015: 2). The selfie becomes a "digital manifestation of a material existence" and the very act of sharing this through a network moves beyond a simple focus on the self.

## 7.5  The Fears of Technology

The study of the selfie and its function in society offers opportunities to consider what it means to be human and how that might be developing or changing as technology changes and develops. This may include particular reconfigurations of the brain and body to enable us to live simultaneously in a material, physical and technologically networked world. Neuroscientists and psychologists in a growing number of scientific studies (Keegan, 2014) have suggested repeated use of digital media may be changing patterns of brain function. Positive changes have included improvements to our problem-solving capacity and complex reasoning skills. The brain is literally re-wiring its neural circuitry through repeated online activity. Negative impacts include distractibility, loss of concentration and impulsiveness. What experts can't seem to agree on is what the long-term benefits or costs might be. For example, do the benefits of an increase to complex reasoning and decision making come at the expense of a loss of more people-centred skills such as empathy? Professor Susan Greenfield (2011) gave a public account of some of these concerns. Attacks on the selfie culture, in some sense, is one of the tangible manifestations of how the underlying anxieties of technology will inevitably change the way humans evolve.

Keegan suggests that although many of the research studies conclude that digital technologies are largely beneficial, it is too early to know what the long-term outcomes may be. All we can conclude with any certainty is that there is a correlation between new technology and changes to the human brain. Adaptability and the human capacity to change has been part and parcel of our survival and optimists would argue the relationship between human and machine is part of the process. We can also see how these anxieties manifest themselves in film, television and popular fiction with stories that attempt to articulate relationships between humans and robots, time and space travel, genetics and biology (see Chapter 9).

The next section continues with the idea that rather than condemn the selfie as an inferior cultural form and the selfie taker as insular and narcissistic we might consider it as part of a communication strategy and at the heart of what makes us human. Technology is just one of the many tools at our disposal in the drive to be social beings.

## 7.6 The Rix Centre and Media Advocacy

Rix Media and The Rix Centre are part of a research centre at the University of East London, UK, led by Director and Professor Andy Minion. They work in partnership with a number of outside agencies including special schools, local authorities, service providers and families in developing multimedia advocacy for people with learning disabilities or who may face challenges in communication. The *Rix Wiki* (http://rixresearchandmedia.org/rix/home-media/, accessed January 2016) was developed at the research centre alongside training workshops in multimedia advocacy and provides an easy to build website for individual pupils to use. The Rix Wiki is described by users as a highly successful tool for "person-centred planning" and helps the individual to have agency and communication in their lives (*ibid.*). The selfie is an important part of constructing a personalised website for people with learning disabilities and a starting place for pupils to build individual goals centred on achievements rather than their limitations. These can be recorded in the form of visual images, sound and voice recording and film clips on the website. Users can decide what aspects of their wiki they want to share with others and what to keep private. For many users it has meant that, for the first time in their lives, they have been able to communicate and have genuine collaboration with adults, such as health and care workers and family members, who play an important role in their care and development.

Taking a selfie is one of the steps an individual can take in defining who they are and how they want to be represented to the outside world. For a person with autism, for example, it is about having a tool to help in communicating ideas and feelings as well as having some control in their interaction with others. It enables people to make and maintain connections and be part of a social world that offers inclusion rather than exclusion. Considering the selfie in the context of the Rix Centre allows us to move far beyond accusations of narcissism aimed at the selfie and see it instead as a visual form with numerous possibilities for social engagement, enrichment and empowerment.

## 7.7  Student Projects and the Selfie

The previous sections in this chapter have indicated that views on how we see ourselves and how we are represented online can become polarised as good or bad in what should be more complex discussions exploring theories of representation, consumerism, data control and protection and gender inequalities. How we use technology to see or reveal parts of ourselves is part of what makes us human. Technology can reflect back to us a version of ourselves that we may not be so familiar with, adding insight and knowledge. In turn we can use technology to create versions of ourselves to share with others (Rettberg, 2014). Digital technology is used to self-document and to record moments in our life and is a way of forging a link with memories and our past. It may form part of a reciprocal conversation or be part of an observed text, read and judge by an audience. In redefining our social media selves, it might also be a way of thinking about how selfies are constructed and the purposes they serve. The Rix Centre Wikis are testament to putting the selfie at the centre of a programme for meaningful social engagement and inclusion.

Students reading this chapter and preparing to produce creative mobile work might consider projects that embrace the selfie as a mode of critical reflection and examine how the selfie is positioned within media debates, both journalistic and academic. You might also consider making work that socially empowers or makes visible hitherto marginalised groups. You have now engaged with some of the key debates and are in a position to counter some of the more negative arguments directed at the selfie if you so choose. Artists, practitioners and individuals are using the selfie as a creative and critical form of expression. This is an opportunity to extend your creative practice and add a new voice to the online community.

Described below are two very different approaches to how some undergraduate student work has utilised the selfie as a way of interrogating the medium and raise further questions. Reflected in both are some of the themes encountered at the start of this chapter. It is important to consider what your own position may be to technology, the selfie, the human condition and society in general. Bear in mind that many of the academic readings try to move away from the more polarised debates found in popular media accounts.

*Me, My Selfie and I* (Nikki Mills, 2014) was a final year student project which engaged with the debates about how young women represent themselves and how multiple personas get created in the negotiation of social media sites. The work consisted of a 2-minute silent film embedded into a still image of a mobile phone. The student explored multiple selfies and performances of the self, which served different functions. Every one of them had a specific audience in mind; the selfie to share with parents, church, or the flirty selfie often found on dating websites. Each image of the self was constructed through a series of layers by the application and then removal of photographic filters. The screen would display all the different styles of representing the self online and how the 'presets' can shape how one is depicted. In this way the constructed nature of the selfie, through the angles and poses and the use of coloured filters, were revealed (see Fig. 7.1). Layering, using the predesigned filters and changing both skin tones and overall enhancement of the photograph was done and undone purposely to generate discussion. It drew on

Fig. 7.1    Me, My Selfie and I — a series of screen grabs from the film by Nikki Mills.

research by Rettberg (2014) who speaks about how filters are an important part of popular visual and technological culture. Using the preset photographic filters on Instagram allow for the boosting of colour, blurring, creating vignettes and borders. However, in doing so, they may filter out the more subtle modes of expression; instant photographic 'fixes' can limit how the self is represented.

Rettenberg discusses how the calibration of photographic film and indeed many digital filters, are biased in favour of white skin tones. The subtleties in black or brown skin and hair are either not captured or lack detail. People of colour have used the selfie (without filters) as a way of reclaiming control of the way they are portrayed in other forms of photography, including professional photographs (*ibid.*). Such images draw attention to the codes of construction at work in popular online photography that become naturalised through constant repetition. They are juxtaposed with the prevailing representation of female identities and provide a space in which to consider how we might use technology as a force for change. This includes considering how commonly held assumptions are embedded in the way particular groups of people are visually represented. In seeking to expose the hitherto hidden power relationships and inequalities that operate as normative functions on the Internet, it enables more empowering uses of technology to emerge.

A second final year student project *Nude* (Georgina Styles, Shaiann Mangan, Anis Davanizadeh, 2015) was informed by some of the media stories on sending selfies with explicit content and how the sender may lose control of the content once it enters the public space of social media. The work took the form of text messaging and images, which gradually revealed the story of a young girl being exploited on a night out with friends. The project was designed with an educational remit to be played in schools to prompt discussions among young people about the boundaries between the public and private, between commonly held assumptions about behaviour and attitudes, and safety online. In this respect, it echoed some of the investigations in Albury's (2015) paper on the use of *sexting* by teenagers.

## 7.8 Conclusion

This chapter has situated the selfie as part of a digital, connected culture. Whilst many of the ideas around self-portraiture and self-representation are nothing new and have developed from a continuum of the need to both look at, be looked at and explore the self, the Internet and social media have re-fashioned how that exploration takes place. Claims by Google suggest that over 30 billion selfies are posted online each year (Mirzoeff, 2015). Much of the media reports surrounding selfies have been seen as largely negative, viewed through the lens of a commodity and celebrity driven culture which sites the individual as a self-absorbed and narcissistic being. Women and young people have been the target for much derision in this respect. By looking at some of the more recent studies that call into question such simplistic readings of the selfie and offer a more nuanced account, we can see the complexity involved in the performance and repositioning of the self in a global visual culture; getting us to think about what it means to be human in a rapidly changing technological landscape. Some of the accounts have also problematised commonly held assumptions about particular groups in society and made more transparent the inequalities operating in the real world that are reproduced online. As such, the study of the selfie offers a space for considering how self-representation might become a tool for self-improvement, self-knowledge and empowerment.

We touch on the celebrity selfie as a distinct genre, perhaps best viewed as a specific category, which shares many features of publicity and commodity advertising. The online self is an extension of the brand and continues to act as part of a marketing strategy where the body and performance of the self are part of the product package. Whilst the non-celebrity selfies may seek to imitate, or even parody those features common in the celebrity selfie, they need to be considered apart when investigating how ordinary people negotiate placing versions of themselves online.

This book centres on practice and I end the chapter by asking students to think about how they might reconsider the selfie in light of the current debates and what the scope might be for widening mobile creative practice. The selfie can be celebrated for the richness and diversity it brings to visual culture whatever the prevailing critiques might have us think. So, students might also consider social engagement projects that connect and re-think the self as part of a more inclusive community. The work at the Rix Centre, which places the selfie as a form of multimedia advocacy for people with learning disabilities, provides us with an example of how the selfie can be truly life changing.

# Chapter 8

# Mobile Journalism: Reflexive Storytelling in the Co-produced Public Sphere

*Max Schleser, Gabriel Moreno Esparza*
*and Anandana Kapur*

Smartphone filmmaking and mobile media has become a regular feature in broadcast news and there are more than a dozen mobile, smartphone and pocket film festivals around the world. Following the accessibility of video production cameras, now post-production is becoming more accessible. Major software companies are producing nonlinear editing software; Adobe, Apple and AVID launched mobile and/or tablet versions for nonlinear editing software. Last year, the British Broadcasting Corporation started offering training courses for journalists in 'mobile journalism' (mojo). Mojo is the current umbrella terminology that includes mobile content creation on the move and encompasses mobile filmmaking, mobile-mentaries (Schleser, 2011), mobile photography (or iphonography) and digital storytelling. As a novel cultural form, this phenomenon is said to have extended the public sphere, not only by opening new venues for ordinary people and alternative voices, but also by paving the way for a more reflexive culture in discursive spaces (Goode, 2009).

In this chapter, we argue that digital stories are a significant influence on accounts of collective resonance by enabling a more mobile and complex demonstration of public impact. To make sense of this we theorise digital video, especially the kind that is produced with mobile phone cameras, as a narrative element that is self-reflexive in nature. Understanding the self-reflexivity of mobile footage is an important step towards making sense of developments in processes of media content production that used to be explained as the by-product of auteurist endeavours, largely ran by editorial gatekeepers, but which are nowadays more fruitfully explained as co-produced assemblages. The assemblages we work with refer to events or happenings that have triggered heated episodes of public debate in India and Mexico after footage segments were uploaded to platforms like YouTube and used by mainstream media to produce news packages. To view these episodes through the lenses that we are suggesting one needs to take several conceptual detours which are required to reflect on the importance of mobile phone video in the nature of contemporary public spheres. While the broadcasting industry functions within the diverse cooperate or public funding models, structures, policies and agendas one can observe how mobile filmmaking appeared as a phenomena internationally and now changes some of the structures within the broadcasting industry internationally. This notion of audience involvement means that face-to-face interactions are brought into a more hierarchical organisation of media dissemination. Glocal impact creates subjectivities and representations within a 21st century discourse. As a cultural practice mojo allows audiences to define their agenda.

## 8.1 Mobile Journalism in the Contemporary Mediascape

The sociological and cultural concept of reflexivity is central to our intervention. The concept mobilised here can be approached as an extension of Anthony Giddens' work on self-identity as the reflexive project of modernity (1991). As he put it, a person's self-identity can be found 'in the capacity to keep a narrative going' (*ibid*.: 54). We can redirect this to Schleser's perspective on self-reflexive mobile filmmaking as a means to 'establishing personal connections and sociability' (2014: 149). Thus, depicted subjectivities and locations become part of 'a public online space' where 'self and life' adopt 'audiovisual form' (*ibid*.: 154). If self-identity amounts to the reflexive understanding of one's own biography as part of a continuous narrative, collective identities are also formed on the basis of shared narratives (Bhabha, 1990). Indeed, as Appadurai (1996) convincingly argued, the widespread availability of new media forms and platforms has triggered alternative ways to imagine social projects, frequently outside the encompassing authority of traditional institutions like the nation-state. Furthermore, these mediated "workings of the imagination" are frequently subject to contestation, and paraphrasing Johnson, become accounts of 'how we struggle, individually or collectively, for some integrity' (1986: 301).

The accounts in question can be redirected to storytelling as an act through which meaning is conveyed. While journalism is one of the dominant practices through which stories are told in contemporary societies (Park, 1940), it is however necessary to consider how mobile media and smartphones are reshaping the way in which stories are told. A new generation of journalists and public communicators are exploring the potential of novel media types and platforms, a process described by Quinn as 'storytelling or presentation convergence' (2005: 6). These are narrative elements formed by video footage, audio clips, timeline infographics and interactive animations, providing not only attractive multimedia environments but also a variety of entry points to suit the interests of multiple users, including the inclination to engage in forum discussions. 'Storytelling is [arguably] social in character' (De Lange, 2010: 174). The interactive potential of storytelling in the new media ecology is also found to be of high significance in the case of what Bradshaw and Rohumaa identify as 'clickable interactives' (2014: 132). This form of storytelling is seen as a potential fix to the ephemeral nature of journalism, supporting long-forms of rather timeless journalism to which people may return for reference reading (Steensen, 2011). Another significant point about storytelling made by Bradshaw and Rohumaa is in relation to the essential role of 'eyewitness footage' for storytelling in journalism, especially when captured by a mobile phone as 'it is often the rawest, unedited coverage that has the most visceral impact on the viewer' (2011: 106). Following aesthetic developments in mobile, smartphone and pocket cameras one can also observe the impact of mobile media on the domain of digital storytelling. These developments are beyond the technological reshaping of storytelling tools found in apps and online platforms, and require investigation as cultural forms if one is to make sense of particular space and places (Schleser, 2014: 154). The emerging locations give way to 'micro-narratives' (Hjorth, 2005). Even the act of uploading said narratives operates as 'a quasi-text, in which symbols provide the rules of interpreting behavior' (Ricoeur, 1984: 58). The 'aliveness' of the event has huge potential for suggestivity, stemming from ontological metaphors of intimacy, natural light, shaky frames and subjective framing of subjects and space invoked by mobile phone video practice also create experiences that persuade and put forth shared experiences. These so-called *aesthetics of mobile media art* (Baker *et al.*, 2009) seemed to be ignored by the broadcasting industry for a lack of technical standard for a decade but are now a defining quality of 'mojo'. With the advancement of video resolution in smartphones personal engagement with stories is unique. As a special feature of video produced with mobile phones, spatial experiences of mobile journalists can be further explored through the affordances of location services (i.e. GPS, afforded by smartphones, GPS as a standard feature in smartphones). Furthermore, even though mobile devices are hyper-location aware, annotation helps transcend geographic places (Tuters and Varnelis, 2011), which facilitates the tracing of people, places and occurrences. Alongside the implicit ethical issues raised by localised 'mojo', one should reconsider how these developments modify traditional paradigms about its nature.

## 8.2 Co-production Acts

In order to make sense of mojo and its acts of digital storytelling one needs to account for the nature of the practice as a co-production. We are informed here by Hudson and Zimmermann's discussion about digital media as 'a process of assemblage' which renders the somewhat invisible and intangible processes of digital documenting as collaborative, reflexive and interactive (2015: 100). Jenkins notion of 'participatory culture' (2006: 3) can be invoked here to recognise that the inputs of the public are essential for the survival of media flows across platforms at the heart of what he terms as convergence culture. Such recognition is present across various sectors of the cultural industries and in general across economic activities (Benkler, 2006). We find this in the way in which news editors and marketers have embraced buzz terms like "citizen journalism" and "interactive advertising" (Deuze, 2005), as they seek to capitalise on the advantages of integrating 'the people formerly known as the audience' (Rosen, 2006) to their business models, and even adapting them to a process of co-creative labour (Banks and Deuze, 2009). The idea is ever present in what Boczkowski labels as 'distributed construction' (2004: 143), whereby users of online resources acquire access to regimes of news production as constant providers of knowledge and information on which newsroom workers rely. In this context, Hill and Lashmar identify the emergence of the 'social journalist' or 'social news gatherer', whose job consists of collecting, authenticating and reproducing content produced by the public (2014: 18). The phenomenon has been significant outside profit-driven environments, as documented by Sambrook's (2005) account of the BBC's reliance on content such as photographs, amateur video, messages and tips from its audiences. The inputs are frequently described as user-generated content (UGC), 'a general term for when non-professionals produce online content' (Hill and Lashmar, 2014: 18). UGC is heavily mediated by the 'social journalist' (*ibid.*), whose job largely requires him to source information from the Internet to produce collaborative journalism 'where journalists work with members of the public to research stories' (*ibid.*). Outside the institutionalised site of the newsroom, Foot (2002) has noted how all political actors, whether citizens or political professionals, become co-producers of the so-called political Web. Professional journalists do indeed understand their jobs depend on their ability to investigate and cover news stories in a collaborative ecosystem (Bradshaw and Rohumaa, 2011: 144–148). Indeed, the tendency of news organisations to 'crowd source' money and manpower to produce, investigate and cover stories has put an end to the traditional idea that journalists are extraordinarily resourceful mavericks (Hill and Lashmar, 2014: 134). During emergencies, calamities and disasters, in particular, the mobile phone establishes a rhizomatic network of eyewitness accounts, alerts and updates. The mobile phone video establishes a community of 'citizen-journalists' that can often provide evidence that affects policy and response.

Whilst one may be sceptical about the reach of the collaborative journalism we describe, its public significance continues to require our attention insofar as the paradigms that link it to democracy (Hampton, 2010) and the public interest continue to bear influence upon the trade (Kovach and Rosenstiel, 2001). Ideas of public impact can be derived from the practice of journalism itself, whereby practitioners have distinguished what it means to produce information that matters to audiences. This may be via "hard news" which has a direct or indirect impact on people's existences or "soft news", described as "the lightweight material that people like to gossip about" (Boyd *et al.*, 2008: 22). Once we assume footage from pocket cameras is a key ingredient of journalism it merges into the type of knowledge which draws the attention of societies to present concerns whilst affording individuals a slow but progressive accommodation into their worlds (Park, 1940). Significantly, news captures the public mind via acts of communication that create social tension or attention. Park considers news affords societies the knowledge to 'understand what is going on about us' (*ibid.*: 672) rather than 'what actually happened'. Since, as Park notes, news is formatted to 'be easily and rapidly comprehended' without significant 'effort of the reporter to interpret the events he reports' (*ibid.*: 670), we could argue the mobile video is similar in that it brings about the experience of the bystander to the rest of the community. Individuals typically react to the news with 'a desire to repeat it to someone. This makes conversation, arouses further comment, and perhaps starts a discussion' (*ibid.*). The ensuing discussions quickly move from the events themselves to the issues these raise, leading to the formation of 'consensus or collective opinion'. At the heart of what the public considers being of relevance lie the sets of values held by communities, which as Anderson (1983) has shown, are frequently imagined across large geographic areas on the basis of shared meanings and experiences.

## 8.3  Digital Narratives from the Global South

The opportunities to bring stories from the everyday life into the newsroom or prime-time news have increased through more and more video footage being in circulation and made available via video streaming sites. The separation of authors' intentions in filming and the framing through institutionalised voice over as much as the juxtaposition of journalistic commentary shows how video footage flows through a number of cultural layers. While a number of news stories originate from PR, press agencies and western newsrooms, mobile media can drive localised or glocal content. India and Mexico as countries with huge mobile and smartphone media uptake suggest themselves for critical analysis. Such is the case because we believe academic discourses about citizen journalism mostly derive from research thrusts in the global north, where accounts on the phenomenon are typically linear. By linearity in this context we refer to the fact that citizen journalists, their platforms and practices such as data gathering, sourcing and authoring are rather unproblematic (i.e. one knows who the citizen journalist is).

Here, the case of amateur footage on video blogs like YouTube as being essential for news storytelling will be addressed to reflect on how factual content is assembled in modes which are collaborative, reflexive and interactive, as opposed to exposed, observed or personalised (cf. Hudson and Zimmermann, 2015: 100). The work of news camera operators and photojournalists has traditionally been represented and understood as the by-product of a creative effort in which individual knowledge, experience, talent and skill are combined to tell stories that matter to an audience. More recently, research has implied the work of digital and/or citizen journalists is the result of individual agency (Holt and Karlsson, 2015; Johnson and St. John III, 2015; Wallace, 2009). One may paraphrase Hudson and Zimmermann to claim that conceptions of storytelling through digital content of audiovisual characteristics are grounded in 'analogue assumptions' (2015: 100). Such assumptions are insufficient to explicate forms of collaboration and interactivity in the production of audiovisual journalistic accounts that are either taken for granted or overlooked. The patterns in question are captured in an online news feature that curated '*5 VIDEOS about discrimination and abuse in Mexico: Of "gentlemen", "ladies" and the bully of a candy-selling child'* (http://www.sinembargo.mx/28-07-2013/700284, sinembargo, 2013). The videos became the object of significant public attention and controversy after they were circulated in social networks and/or edited by broadcasters as digital packages. The feature draws on footage taken from various sources and posted on YouTube between August 2011 and July 2013. The spread of the dates suggests the clips are but a sample of a growing collection of audiovisual segments on YouTube which illustrate a societal culture of power abuse. Such a culture has in recent years become a recurrent trending topic tagged with the English-language words "Lady" or "Gentlemen", which for Mexicans invoke the idea of refinement and decor. The terms have been used sarcastically in reference to the exposed corrupt nature of individuals who abuse positions of power acquired from occupying jobs in high public office or from being famous and well connected.

Due to the emphasis of the videos in question on abuse of power, the trend could be thought in relation to ideas of citizen journalism documented by various authors (Allan, 2007; Sambrook, 2005). Notably, the videos described above were said to have caused some form of public outrage and in every case, the individuals involved had to face justice in court, paid fines or were removed from their jobs. This could be taken to be an extension of the now popular notion that journalism sheds light on power abuse and injustice, frequently acting as the first draft of history. In the age of mobile phones and vblogs, these first drafts are the result of patchy strands of collaborative endeavours rather than individual journalistic doggedness. Authorship is no longer the repository of a single creator and narrative arcs emerge via metadata — #hashtags, hyperlinks, comments, ratings and shares. The exposition, reproduction and even re-appropriation also create crises of authenticity and origin. 'There is a post-modern paradox of authorship. Real people are censored and anonymous people have a right to say everything' (Subrahmanyam, 2015: para 7). There is a conceptual (and productive) blur between affect-driven infrastructures and the movement of media. Ficto-graphic atrocity stories (images, sounds, videos) circulate and attach themselves to sites of violence; in India, for instance 'fake' videos have been held out as reasons for disturbances in various cities and for the intimidation and killing of minority populations (Sundaram, 2015: para 16).

The first of the videos (https://www.youtube.com/watch?v= FhGcqVfaeLg) shows infamous entrepreneur Miguel Sacal beating and verbally abusing Hugo Vega, a receptionist at Sacal's luxury residential building in Mexico City. The attack started after Vega refused the demand that he changed a flat tire in Sacal's vehicle, which to start with was not part of Vega's job duties. The footage, which was widely reported by national broadcasters and the press, was entitled 'Abuse and violence', uploaded by a YouTube user identified as 'suy020204' who claimed to have retrieved it from an unidentified human rights group. The footage was edited with dramatic music in the background, and included halo effects for emphasis on each of the participants; it also included subtitles that provided both narration and interpretation of the events, such as: "Sacal leaves his jacket on the desk and walks towards Vega to beat him. He also insults and humiliates him."

The second video featured was entitled *Las Ladies de Polanco* (https://www.youtube.com/watch?v=6QCrGKm8mg4), which shows a former Big Brother contestant and a female friend in a wealthy part of Mexico City swearing at a policeman and calling him "working-class piece of shit".[1] The evidence was widely commented by Mexican newscasters and social media users as evidence of uncivil behaviour from a foul-mouthed second-tier celebrity, but it also generated debate about the failure of the police to inspire the respect of citizens. Here again, the source of the video was sketchy; the video had originally

---

[1] A similar abusive refrain was captured by an Uber customer Juan Cinco who uploaded a video (https://www.youtube.com/watch?v=Bvq07KBfhnQ&feature=yo utu.be, accessed March 10, 2016) of a visibly inebriated Dr Anjali N. Ramkissoon attacking and swearing at an Uber driver on January 19, 2016 in Miami. The video went viral soon after it was uploaded leading to outrage and trolling of Ramkissoon, including cyber-abuse such as morphing of her photographs. While Ramkissoon has been placed on administrative leave and has pulled down all her social media accounts, the assemblage of comments, embedding and sharing will continue to imbue the event with greater aliveness.

been uploaded to a YouTube account and then changed to a different account. The footage was framed in a narrow vertical shot and whilst it was taken at night it clearly captured the threatening body and verbal language of the women, as well as the indistinct conversation of the individual shooting the video with other witnesses. At one point, one of the police's assailants told the person with the mobile to make sure he'd get a clear shot with the face of the policeman, who appeared embarrassed through the 84 seconds of the recorded fracas.

The third video referenced shows footage of a drunken man Francisco Arias at a police station. The private assistant to the major of Oaxaca city, he repeatedly abuses and challenges a municipal cop to a fight. The footage was part of a piece presented by a news reader and broadcast on pay TV by the national news media group Milenio, which itself uploaded the edited segment to YouTube. The package included subtitles announcing: "Major's PA fired", followed by information noting the location of the story and that "footage was released in which he (Arias) verbally abuses cops". The "footage was released" bit in the subtitle suggests the video was not produced by Milenio's journalists and that it was indeed passed on to the newsroom by someone at the premises of the police in Oaxaca where the events took place. By watching the video, in which Arias slaps his left biceps warning a circumspect cop how he would suffer from one of his punches, one can further speculate that it was taken with a phone camera, as the good visibility of close range actions turned grainy when the subjects became slightly farther from it.

In the fourth video (https://www.youtube.com/watch?v=0NS5f9lqrKE) Francisco Romo, a local councillor of Ixtapaluca County is shown at a street food stand under the influence of alcohol proffering abuse to a policeman as he demands that he submits to his authority. The policeman, who was with some of his colleagues who appeared in the footage, remained calm recording with a mobile camera as Romo swore and gesticulated at him. At some point the former questioned the latter's authority suggesting his erratic behaviour would leave him exposed. It's worth noting that this piece of footage was uploaded in raw format by national newscaster MVS Noticias, which simply used elements of information within the video itself to write the title on YouTube: "I give the orders, respect those above you", says member of Conservative party, aka "the Ixtapaluca gentleman". This suggests MVS Noticias gave little if any attention to who recorded the video or the question of who was captured in it, an aspect which was nonetheless specified in the feature by the news site which assembled the five videos in question.

The same Sin Embargo feature further documented the specifics of the fifth video, (https://www.youtube.com/watch?v=IxVv9w_XlBw) allegedly made public on Twitter by a local lawmaker and uploaded to YouTube by Canal TVX, a broadcaster from El Salvador. The footage, which was apparently produced by Canal TVX, captured the moment in which local trade inspector of Villahermosa (capital of Mexico's southern state of Tabasco) Diego López allegedly 'humiliated' Manuel Díaz, a kid from an indigenous community. In the footage, Diego is shown holding a straw basket from its handle as little Manuel cries whilst emptying the basket of candies he was trying to sell in the street. Once he emptied the basket Díaz is seen leaving the scene, after which a man is seen putting Manuel's products back in the basket whilst the child squats and cries with his head sunk between his knees.

## 8.4 Of Gender Violence, Injustice and Corporal Punishment

Eponymously named after the duo filmed in a mobile phone video that went viral,[2] the 'Rohtak sisters' or 'Sonepat sisters' video (https://www.youtube.com/watch?v=ZopUkgU7CSw, accessed March 2016) was uploaded on November 28, 2014 and subsequently embedded in news reports with anchor links and voice over. The footage shows two young girls retaliating against a group of 'molesters' in

---

[2]A headline that is typical in mobile videos is the phrase 'video goes viral'. One can assume that the act of sharing is indexically applied to convey consumption as act of public significance rather than simply as a process of mass communication. Thus, these cases allow us to conceptualise processes of co-production in digital networks as sites for the enactments of counter-abuse.

a bus in India. The physical scuffle between the two parties is captured clearly in the footage. After the video was uploaded journalists,
women's commissions and politicians lauded the bravery of the girls.
The police rounded up the alleged molesters soon after and while the
investigation was on another video (https://www.youtube.com/
watch?v=GkA77v8Y4oc, accessed March 2016) went viral a few days
later. This showed the same duo in physical conflict with another
group of boys. Were these 'serial bravehearts' or 'bullies' asked the
press and online users. What was the footage an index of? On
December 4, 2014 yet a third video (https://www.youtube.com/
watch?v=Vgu3ibsHwDI, accessed March 2016) in this saga was
released which showed the boys from the bus being chided and disci-
plined by their parents. With the latter two videos going viral, public
opinion turned against the sisters. As if on cue, several female witnesses emerged, stating the sisters had misrepresented a dispute over
a seat as a case of harassment by releasing the first video. Questions
were raised about the first video being 'unedited' thereby undermin-
ing its credibility as a witness account. Any critique of the passivity[3] of
the mobile cameraperson during the altercation is turned on its head
when the woman who recorded the event testified that she began the
recording at the behest of the two sisters *after* they had argued with
the man about the seat. The sisters volunteered to undergo a poly-
graph test to silence all counterclaims, but failed it. The plot took a
further turn when morphed images of both sisters began circulating
on social media and they in turn had to file a complaint with the

---

[3] In a road accident in New Delhi in early 2016, 32-year-old Manik Gaur was charred
to death when his motorbike collided with a stationary tempo. Onlookers shot the
gruesome incident while the biker fought for his life. The apathy of the onlookers and
the fact that mobile videos were uploaded by several of them under 'entertainment'
sections of video platforms online evoked a sharp reaction from the deceased's family.
They objected to the apathy of the witnesses and the possible trauma that circulation
of the footage would cause to them and are pursuing the case with the Cyber Crime
unit in New Delhi. The incident clearly highlights the ethical implications of witness
footage and the complicity of the video makers in the outcomes of the event
being filmed.

cyber-crime cell of the police. This provides an interesting insight into public notions of social injustice.

Another stamp of co-produced content in India stems from a September 10, 2015 mobile phone video in which a teacher beating a 10-year-old boy with a fractured arm went viral. The teacher is seen constantly haranguing the boy about his caste and brutalising him. Though the incident occurred in April of 2015, the footage was released many months later and led to a police complaint by a Dalit[4] organisation. The conduct of the teacher would have gone unnoticed had it not been for the anonymous onlooker who filmed the act and uploaded it. It is not known, whether the teacher was aware of being filmed but the episode unleashed public anger, triggering calls to address what was perceived as an act of impunity and Twitter hashtags such as #everythingwrongwithIndia. Both corporal punishment and caste-based abuse are not uncommon despite both being punishable offences in the country. The footage is an important testament to the power of the mobile phone video as a catalyst for political and social justice, albeit belated. The video, originally shot in Mangalore, trended immediately after its release but has been removed for alleged violation of YouTube's community guidelines. Grainy screen grabs remain in online news articles whose two-dimensionality can never do justice to emotional and psychological dimensions of the incident. The absence of the complete video and the conscientious comment by the corporate video aggregator play up the dialectic of how these co-productions re-enact the role of journalism as they seek to break silences, whilst subject to traditional editorial processes practiced within mainstream media organisations. However, the multiplier

---

[4] Dalit refers to a caste (not a class) in India that were discriminated against and treated as untouchables because of their occupations (manual scavenging, leather work, butchery, etc.). Meaning 'broken', 'downtrodden' or 'oppressed' the nomenclature Dalit is an act of political self-assertion by those communities that were isolated, discriminated and subject to violence by the rest of society. Even though the caste system has been abolished in India, cases of exclusion and discrimination continue to exist. Pervasive issues affecting Dalits include debt bondage, low levels of literacy, exploitation and impunity of perpetrators of crimes against them.

effect of such videos can be seen as part of automatic playlists where students are mercilessly being subjected to corporal punishment.

Embedded in TV news reports, videos from pocket mobiles are used as continuously used as commentary on the need to challenge social tolerance of physical and emotional violence. Another such video(https://www.youtube.com/watch?v=EWsDJoNiiVc&feature=youtu.be, accessed March 2016) emerged in May 2015, showing a college principle caning his students as well as reactions of students who witnessed the event. The naturalisation of violence is evident when a student, over whose shoulders the mobile video is being recorded, turns with a grin but immediately stops on realising that his position as an approving spectator may become irrefutable. A little further in the video, another student realises that the video is being made and tries to clear the line of sight of the mobile camera. He and the other student clearly recognise the implications of the mobile video being created. A little later, one of them even pulls out his own mobile phone, as if inspired to film from the front row. This performance of complicity with the anonymous student who filmed the event is a demonstration of how *mojo* transgresses and redresses boundaries. In this case, a classroom as a site of violence becomes released from the fear and hierarchy through its visibility. In a news-report, in which the video is embedded, the creator asserts that silence was not an option. The emphatic insistence of the police on verifying the veracity of the clip is however ironically directed at investigating whether it was doctored and not. The inability to convert witness footage to evidence in a judicial context points to the limitations of such footage. The purpose and outcomes of *mojo* may not converge in many instances, but in this case the entire student community decided to file a case *en masse* so as to not jeopardise the anonymity of the creator of the mobile video. Since both videos of corporal punishment originated in the same region (southern state of Karnataka), one could speculate with the possibility that the creation of one was inspired by its May predecessor.

As demonstrated by the students' coordinated strategy it is clear that actions in the offline world can be triggered by *mojo*. Thus, lack of consensus on the events that took place on the bus ride and questions about the authenticity of video can be overlooked in favour of under-standing the political implications of this generalised condition of mobile video content in circulation. This can be redirected to the notion that 'With the cellular phone, a growing section of the postco-lonial population is now the source of new-media output — which in turn links to online social networks, mainstream television, and peer-to-peer exchanges of text, music, and video' (Sundaram, 2015: para 10). Mobile phones are also repositories of 'found footage' (i.e. videos that have the potential to be circulated and carefully curated in terms of duration and point of release). Those who put, capture or circulate the footage in question, in fact 'expose the failings of public and private institutions and their personnel, and sometimes become celebrated opinion-leaders' (Goode, 2009: 1290). At the same time, 'sense mak-ing' of the videos is shaped by traditional gate-keepers and agenda shapers. Academic work too has raised serious objections[5] about the significance of the input from non-professional journalists (Buckingham, 2009; Wardle, 2010). We should therefore consider some of the prob-lems with the nature of the videos where the label 'citizen journalism' could be more accurately replaced with participatory (Henig, 2005), witness (Wallace, 2009) or random (Holt and Karlsson, 2014) journal-ism, particularly when it comes to questions such as the role of the author in mobile video. Because it is common to deal with footage that has no indication of having been obtained with typical journalistic pur-poses, one could at best argue for the agency of 'witness contributors' (cf. Buckingham, 2009: 98). In the case of the videos discussed the question of who shot the videos is highly problematic in that there are little if any clues as to the identity of the camera operators. This raises several problems, including a somewhat historical tendency to consider the work of professional camerapersons as somewhat secondary, and therefore unaccredited, to that of news presenters and reporters. One could of course raise questions as to the editorial professionalism of

---

[5] See Brennen, Bonnie (2010) "Photojournalism: Historical Dimensions to Contemporary Debates", Stuart Allan (ed.) *The Routledge Companion to News and Journalism*. Oxon; New York: Routledge, pp. 71–81 for a discussion on the impact of digital photography on authenticity claims made by photojournalists and documentarists.

mainstream news organisations that simply chose to acknowledge the source of content from witness contributors or more problematically, to provide any evidence that measures were taken to ensure the content was reliable in the first place. There is then the fact that well before the emergence of mobile cameras news was being packaged with 'found footage' from CCTV facilitated to news organisations by institutional sources who were key providers of content for OOVs (short for "out of vision" digital packages) (Hill and Lashmar, 2014: 105). At present, the distinction between video from participatory interventions and surveillance cameras is short of straightforward. Certainly, footage from CCTV generally provides a soundless *fly on the wall*[6] sort of perspective whilst the one obtained with a phone camera will be normally level to the ground and with some form of audio. But the final result will be very similar in that footage will tend to feature grainy, poorly lit, with reduced frames, occasionally blurry but good enough as evidence to support the claim made by professional journalists in their digital packages.

## 8.5 Mobile Media as an Agent for Change in the Digital Public Sphere

While mobile camera phones were never intended for filmmaking when they first appeared, filmmakers and creatives defined the aesthetics and working practices. The beauty of mobile filmmaking is exposed through the creative exploration of filmmaking and its break from established rules and conventions. Writing in *Spectator, Film and the Mobile Phone*, Roger Odin argues that cinema is no longer only a

---

[6]The concept of direct cinema and its fly on the wall approach paved the way for our understanding of contemporary TV and cinema formats. Beyond the recognition of industry formats it is particular interesting for vernacular creativity and formats that emerged out of audience interaction. A number of scholars have critically engaged with this concept and this recognition underpins this chapter. See Winston, Brian (1995) *Claiming the Real: The Griersonian Documentary and Its Legitimations*. London: BFI Publishing; Winston, Brian (2000) *Lies, Damn Lies and Documentaries*. London: BFI Publishing and Winston; Brian (1996) *Technologies of Seeing: Photography, Cinematography and Television*. London: BFI Publishing.

matter of films but has become a language of communication; '…
today film language, when it is not used to make 'cinema' is no longer
confined to certain areas of specialised communication, but is mobilised by the space of everyday communication. The era of film language
has truly arrived' (Odin in Christie, 2012: 169). Access to smartphone
technology has meant that more works[7] are surfacing from countries
beyond western screens and broadcast networks. In combination with
the opportunities to disseminate videos great opportunities for more
diverse content from a number of countries emerges. The next generation of filmmakers will utilise the mobile device according to their
own ideals and agendas. Mobile filmmaking and MoJo is engaged in
a constant innovation process that is influenced by multiple vectors.
It is emerging as a field with its own aesthetic qualities. Mobile films
capture a space that is often overlooked — revealing moments of life,
capturing the mundane in a poetic way reminiscent of the essay film.
Whether waiting at a bus stop, inline or going through a car wash, the
smartphone is always with us. Our thoughts in these moments, which
allow us to reflect upon the current moment and live is captured using
mobile visual media.

While the editing process allows filmmakers, creatives or journalists to reflect upon the captured action and craft this into a story
format, user-generated content is normally understood as a form of
video that has minimal or no editing. As we have shown with the
above outlined examples, the editing is replaced by discursive forces
in these witness accounts. Mobile video and especially *mojo* is thus
shaped by an institutional voice-over. Historical precedents such as
Zapruder 8mm footage or the VHS of the Rodney King incident have

---

[7]The Mobile Innovation Network Australasia [MINA] aims to explore the possibilities of interaction between people, content and the creative industries. The annual
*International Mobile Innovation Screening* and the *Mobile Creativity and Mobile
Innovation Symposium* have become widely recognised for the debates within and
beyond the fields of media, art and design. In its fifth edition, MINA is continuing
to grow as a network project between the College of Creative Arts (Massey
University, NZ), Co-Lab (AUT University, NZ) and now RMIT University. The
2015 MINA submissions were more than those in the last 4 years put together. Most
submissions were received from the USA, followed by Iran and India.

demonstrated (Nichols in Renov, 2010) the context of video footage and its placement into sociocultural contexts. Besides the form of news, celebrity and yes also 'cat videos' some forms of live streaming (Ustream, Mini WebCam, TwitCasting or Bambuser) can be referenced to the form of actuality videos. The well known and theorised *L'arrivée d'un train en gare de La Ciotat* (Auguste and Louis Lumière, 1896) or *La Sortie des usines Lumière à Lyon* (Auguste and Louis Lumière, 1895) mirrors contemporary micro-formats of Vine, Instagram Videos or Vyclone. The notion of mobility is characterised by its detachment from desktop production, linear distribution and a more user-focused approach to film exhibition. These transformations are increasingly driven by a horizontal media ecology and dynamic mediascape. Messages are personal, intimate and immediate. Some of these works are reflections on life and others on art and culture. Their meanings are powerful as we can connect to the thoughts of the filmmakers. Some works tackle social problems through a strong statement while others allow us to understand situations of people and their lives.

Before we look into the future and make futuristic predictions we can look back to the properly first ever *mojo* producer and digital storyteller Dziga Vertov.

'Freed from the boundaries of time and space, I co-ordinate any and all points of the universe, wherever I want them to be. My way leads towards the creation of a fresh perception of the world. Thus I explain in a new way the world unknown to you.' (1927: 18)

By means of exploring reflexive storytelling to theorise mobile filmmaking we demonstrated some considerations that need to be attached to *mojos* shaping as a format. As global phenomena we choose examples of media stories from India and Mexico as one can observe how mobile media now contributes to the development of digital public sphere. The Internet's potential for non-hierarchical dissemination of content through peer-to-peer networks obviously opposes the industries structures. While *mojo* can emerge from audiences in a grass roots bottom-up fashion, the dissemination is flowing through a broadcasting model that conceptually did not change notions of dissemination since its proliferation in the 1950s and 1960s.

The 'alterity' of the narratives emerging from mobile phone videos lends them plurivocality. Akin to magical realism, the digital mode is also '...suited to exploring...and transgressing...boundaries, whether the boundaries are ontological, political, geographical, or generic' (Zamora and Faris, 1995: 5). 'It is [therefore] necessary to research and seriously debate the extent to which a culture of — or demand for — "reflexive conversation" matches up to the potential evident in the online news sphere' (Goode, 2009: 1302). Till then we can collectively say — thankfully we have the mobile phone video to frame what happened.

# Chapter 9

# A Question of Ethics?

'Socrates, the 'father' of philosophy, proposed that people will naturally do 'good' if they know what is right. But this point might only lead us to yet another question: *how do we know what is right?*' (Fox and Caruana, 2012: 171).

This final chapter of the book investigates the complexities of creating ethnographic, documentary and photography projects where the human subject is the object of analysis. It will examine particular case studies by way of illustration and encourage students to focus on their responsibilities as photographers and creators of media. Creative mobile media and the use of smartphones to create project work can occasionally give rise to specific ethical questions related to the technology but, by and large, when considering how to work ethically we are considering the framework for all kinds of practice. This chapter therefore engages with the wider questions regarding ethics that inform what we do. Ethical considerations encompass questions around choice, values and morality. A starting place is to consider some ideas around these terms and their usage and how this may be relevant when thinking about the construction and reception of the creative work being produced. We consider the responsibilities we, as practitioners/producers of work for a public facing audience, have to the human subjects at the centre of any of our projects. When we speak about ethics and working within an ethical framework, it is important to consider what exactly that means in terms of practice and our everyday encounters.

Some of the earliest philosophical debates on ethics were shaped in ancient Greece. The study of ethics is a branch of philosophy concerned with how we live our lives in respect to other humans (and indeed, non-humans) and in recent times has expanded to cover how what we do and the choices we make can impact on our planet. Questions regarding ethics, therefore, are concerned with sets of values, value-judgements and moral choices that humans make and their affects and consequences.

Moral philosophy or ethics considers particular codes of conduct, how we behave in society, what laws or rules are involved to enforce certain kinds of behaviour and questions of 'right' and 'wrong.' Values are a set of beliefs which are culturally determined and have both shared meanings and a personal investment attached to them. They are a set of ideas that that are personally important to us and help us organise our behaviour. They are not fixed and may change through time although for many people they can remain relatively stable. Values often get confused with facts, which can be defined as objective or bias free, evidence-based knowledge. The difficulty in making clear distinctions between the two highlights some of the complexities in any discussion on morality. What appears to be factually based may in fact be the result of commonly held assumptions that through time get normalised as part of the 'social landscape' (Smith, 1998). One of the purposes of ethics, therefore, is to consider, question and challenge where ideas come from and the basis of 'factual' information. It enables us to draw distinctions between common practices, sets of beliefs and more evidence-based knowledge. In turn, it gets us thinking about knowledge not as a given but in terms of its construction and how it is authenticated. This will be considered further when we start to look at work that makes claims to have a particular relationship to objectivity and truth.

This chapter will continue by looking at popular culture and some of the ethical issues raised by particular modes of media production. It concludes by an examination of how the Internet and the digital world's infiltration into all aspects of life, present particular issues around communication, ownership, copyright and privacy.

## 9.1 Ethics and Popular Culture

### (a) *Folk and Fairy Tales*

Some of the earliest forms of stories we encounter as children are in the form of folk or fairy tales. These have been described as morality tales and are understood as stories that carry a message. Indeed, Jack Zipes, in the preface to his book *The Trials and Tribulations of Little Red Riding Hood* goes so far as to say the literary fairy tale was 'consciously cultivated and employed in seventeenth century France to reinforce the regulation of sexuality in modern Europe' (1993: xi). By tracing the incarnations of a story like Red Riding Hood, Zipes argues shifts in attitudes to gender, relationships, sexuality and the use of power can be tracked. It is in such stories that many current social norms and discourse on manners have their roots. In this way, it might be argued, some of our earliest experiences of grappling with moral questions and ethical dilemmas have been encountered, in the first instance, as children in the stories we heard and read.

In Part One of this book, we examined some of the basic building blocks for constructing stories and considered how the fairy tale had been unpicked in terms of themes and form. The writings of Vladimir Propp and Tzvetan Todorov have been a key influence on generations of filmmakers and writers for their ideas around narrative structures. Inherent within such narratives were encoded particular rules of how to live; good people were rewarded, those who transgressed were punished or put to death. Folk and fairy tales have their roots in the oral traditions of storytelling and many tales familiar to UK children have European roots. Jacob and Wilhelm Grimm have been credited with the written versions of many of the western European fairy tales, popularised in the 19th century and which, by way of the Walt Disney Company, now have a global reach. Many countries have variations on the basic ideas and themes within these stories whilst having elements that are geographically and culturally specific. The continents of Africa, Asia and Central America have a rich history of folk tales in which the specific differences and similarities to the western tales are fertile grounds for exploration and can illuminate some of shared and different cultural practices. India is thought to be the source of many western tales that have crossed borders and have been adapted over time. The collection of globally known stories popularised in the *Arabian Nights* was thought to have their origins in the Middle East and India. Scholars have suggested that the nomadic nature of oral storytelling give rise to motifs and plotlines from the folktale travelling the world. Other theories suggest that archetypes and the common experiences of human society may have resulted in the similarity of narratives rather than as a direct result of contact through travel (Warner, 1994). Propp's (1927) research was based on Russian folktales, however, he identified certain basic themes and structures which appeared to be common to all tales, that is, they had some universal themes and values.

The folk and fairy tale introduce us to a range of stock characters. Differing from modern literature with its emphasis on complex characters, backgrounds and sub-plots, the focus was on the action and what happened to them and what comes next (Pullman, 2013). Propp identified these in terms of the functions they perform and the attributes they embody. They included the hero, basically a good character who helps to protect both individuals (in the story) and by extension society at large; the villain, a bad/evil character who threatens the stability of society; the false hero who gives the impression of being good but ultimately is shown to be on the side of the villain. There is usually a 'love interest' in many of the tales, a princess and winning her hand in marriage is a popular reward for the hero. Needless to say, many of the stories are set within a patriarchal hierarchy with the actions of the female characters following stereotypical and limited routes. Feminist readings of folk and fairy tales have, however, widened the debates around gender by investigating the ways women, who were often the storytellers, have subverted traditional roles and plot lines and made space within the narrative for more nuanced readings (Warner, 1994: 182). Others *re-tell* the tales within a feminist framework that seeks to highlight the myriad ways women are resourceful even in times that have sought to silence their industry and thought (Carter, 1979).

In a Proppean analysis, each character served a particular function within the narrative and was part of a set of binary opposites, which served to reinforce the functions and actions of each character. There is little in the way of complexity or subtlety as morality tales serve as warnings against transgressions in behaviour. Universal lessons in greed, murder, cruelty, lust and marital relations are common to many of the stories but by uncovering particular histories more is revealed into the changing and material circumstances of the time. For example, the changes in women's roles in the more traditional tales can be compared with some of the more recent re-writings, particularly by female authors and scholars. Some of these interpretations have sought to give many of the attributes of the hero figure to female characters, whilst the much maligned crone or witch has taken on the mantle of a 'wise woman' (see Warner, 1994: 182; Carter, 1979).

Feminist readings of the tales have highlighted not only the inequalities in gender relations but also the changing mores in the control and containment of women's sexual behaviour. The traditional story of *Little Red Riding Hood* is usually regarded as a cautionary tale for young girls; if they 'stray too far from the path', they are likely to get devoured by the 'wolf'. In some readings of the tale the red cloak of Red Riding Hood has been evoked symbolically to represent the first menstruation and the transition from childhood, and that a girl's honour is thus tied to her conduct. The warnings to Red Riding Hood 'not to stray' both literally and metaphorically are passed on by her mother and it is made clear in the story the consequences of transgressing the societal rules governing a girls sexual behaviour will result in punishment. In many of the tales she is de-flowered/devoured by the wolf and in others she can only be rescued by the wood cutter/father figure. A particular morality and code of behaviour based on gender is thus encoded within such stories. In the coda to the story translated by the 17th century writer Charles Perrault from his work: *Tales from Times Past with Morals* or *Stories from Mother Goose*, Warner comments on how the wolf figure is updated from the wild and uncivilised creature of the forest to a more modern, urbane character.

> 'Now, there are real wolves, with hairy pelts and enormous teeth; but also wolves who seem perfectly charming, sweet-natured and obliging, who pursue young girls in the street and pay them the most flattering attentions. Unfortunately, these smooth-tongued,

smooth-pelted wolves are the most dangerous beasts of all.' (Warner, 1984: 182)

Angela Carter's re-telling of the story, which was subsequently made into the feature film *Company of Wolves* (1984, UK), plots further the 20th century's shifting landscape in thinking echoed in feminist culture and writing. The Red Riding Hood of Carter's story is symbolic of a more open attitude and tolerance in more liberal societies to the recognition of women with agency. Red Riding Hood willingly goes into the arms of the wolf and the balance between male and female desires are equivocal. If the wolf is skilled in the art of seduction, so too is Red Riding Hood, who very quickly learns the nature of her own desires.

## (b)  *Film and TV Fiction*

From myths and folk stories to popular films and television drama the depiction of characters and development of storylines present a myriad ways to consider questions on ethics in popular culture. The American Western and James Bond films have both undergone Proppean analysis by academics (Wright, 1975; Hollows and Jancovich, 1995). The classic western of the 1940s with its hero (usually wearing the white hat) as the stranger who comes to town, defeats the villain/s and returns society to the rule of law, forms the basic Proppean structure. Good is seen ultimately to triumph over evil and the hero gets his reward in the shape of the available love interest. Punishments or death awaits the villains. These symbolic stories are worthy of study as they engaged with both continuities and changing social practices and alert us to the shifts in cultural values. Some academics have also indicated they uncover particular ideologies at work in creating powerful myths that function in maintaining social norms (Monaco, 2000). Mainstream Hollywood cinema and more recently TV dramas, allow us to trace the development and engaging complexity in how 'goodies' and 'baddies' are represented and what constitutes these identities. In the 1960s, Russians were the antagonists of many thrillers reflecting a preoccupation with cold war politics. Some current narratives in American drama offer a generic middle-eastern terrorist as the key villain as, for example, depicted in the TV series *Twenty-Four* (2001, USA). and *Homeland* (2011, USA). Whilst subject matter, narrative and characterisation may well be a reflection of social concerns, anxieties and even western ideologies, some of the most recent media and TV outputs have tried to capture a more nuanced, sophisticated and complex depiction of the world. Rather than present ethics and morality as simple choices between right and wrong in a framework of clearly delineated good and bad behaviour, they engage with the contradictions and ambiguity of the human condition.

*The Wire* (2002–2008, HBO, USA) was a critically acclaimed American TV drama that told the story of the city of Baltimore and unfolded in the course of five seasons. It has been likened to a novel in terms of its structure and complexity with each series examining a different aspect of the city, its institutions and citizens. The storyline for Series One was based around Baltimore's inner city drugs problem and focused largely around the characters of the drug dealers, drug users and the police. What made the series exceptional, when we consider questions around ethics, was its refusal to present simplistic notions of good or bad, right or wrong. Angela Anderson writing in the online journal *Dark Matter* comments: 'In *The Wire* there is no such thing as good and evil as clear-cut moral categories. There is no heroine or hero, but a cast of at least thirty significant central characters' (2009: online).

*The Wire* was also credited with producing more authentic and nuanced black identities. This is a significant change from the more usual, simplistic tendency to utilise character stereotypes. On character roles, Anderson states: 'For the most part young, inner city black men are transformed into living breathing people who actively struggle with questions of ethics as well as survival' (*ibid.*). The police are seen as dysfunctional individuals, government officials are corrupt, drug dealers show acts of kindness and humanity alongside possessing a keen business sense. This isn't just a reversal of the sanitised and traditional power structures usually depicted in some mainstream TV drama but an attempt to create characters and situations that depict realistically and convincingly the complexity of life. In this way the city of Baltimore, its citizens and institutions though still a screen representation and dramatisation reflects, in a more realistic way, the experiences of a real-life American city. There are no clear-cut 'good' or 'bad' characters or a single hero or heroine as found in fairy tales. We see the ambiguities, contradictions and moral dilemmas of life that are recognisable to an audience. We might consider, therefore, how ethics plays a role in shaping characters and their actions in these latest screen outputs.

What makes *The Wire* worthy of examination *vis-à-vi*s ethics is that it operates on two levels. Firstly, it tries to portray a more complex and realistic reading of identity in the way it represents people of colour, gender and sexuality. Furthermore, there are questions to be asked about representation; how people have and are being depicted and who gets written out of the picture. Secondly, beyond the world of the story and to the wider responsibilities of the cultural industries, it has provided many credible roles for black, Asian, minority ethnic actors in an industry where historically, these have been limited.

Our next case study in this section examines the TV series *Humans,* first screened on Channel 4, UK in 2015. The series was described by Channel 4 as their most successful drama for 20 years and was watched by an average of 4.8 million viewers. Produced by *Kudos Film & Television,* it was a UK adaption of a Swedish drama series by *Matador Films* titled *Being Human* (Humans: Channel 4 online, 2015). The story takes place in a parallel present where it is perfectly normal to have a *Synth* — a synthetic, lifelike humanoid — in one's home to help with the domestic chores. As the storyline unfolds the family who have acquired their first Synth at the start of the series, gradually realise it is not behaving as it should. The Synth has moved beyond being simply a machine and has a conscience along with the full range of human emotions. It belongs to a small band of specially created hybrids, a family whose members have become separated from each other and who spend the series trying to reunite. The human family gradually start to become emotionally attached to the Synth whom they have named Anita with the teenage son even developing an adolescent crush on her. However, the elder daughter,

Mattie, is antagonistic towards the Synth. Machines and Synths have replaced humans in much of the job market and Mattie and her friends see no future for themselves. To make matters worse, Mattie's father, by the fourth episode, has switched on the 18 plus adult setting on the Synth to have illicit sex with Anita whilst his wife is away.

Humanoids, robots and artificial intelligence are not new to literature and screen-works. *Blade Runner* (1982, USA) was one of the most successful films to present a future in which artificial intelligence (AI), in the form of fully sentient humanoid beings, were part of the landscape. Not only were they utilised in all aspects of work and the home environment, pleasure models were programmed to work specifically within the leisure and sex industries. *Cloud Atlas* (2012, USA) based on the novel by David Mitchell had a scene set in the future in which a lifelike robot used for menial tasks and sex rebelled and went in search of a better life. Since digitisation, however, there have been rapid advances in technology and what was conceived as a fictional and futuristic world is now becoming a reality. The drama *Humans* was transmitted by Channel 4 at a time when artificial intelligence was a major news story. What is manifest throughout the narratives and subplots of the series, and which probably added to its popularity, are the fears and ethical debates currently engaging the real world. Whilst experts have declared such lifelike creations are some time away, robotics and AI have, nevertheless, developed apace. They are part of the technical landscape with programmed machines used routinely in hospital, education and manufacturing. In Japan, *Erica* developed by Hiroshi Ishiguro at Osaka University was presented as one of the most human-like robots able to converse in around 20 subjects. The robot was designed to simulate human expressions and mannerisms and was part of the Symbiotic Human-Robot Project (Ishiguro, 2015).

In December, 2014 Professor Stephen Hawking spoke to the press with a warning that: 'full artificial intelligence could spell the end of the human race' (Cellan-Jones, 2014). Other expert in AI and robotics are more optimistic about what the future may hold and how human lives might be improved as a consequence of these developments. Author David Levy (*Love and Sex with Robots: The Evolution of Human–Robot Relationships*, 2007) predicts relationships between humans and robots will be 'commonplace by 2050' and that there is a market to fill the void of those who have difficulty in forming relationships    (http://www.bbc.co.uk/new/technology-34118482, accessed October 2015). His book offers up the possibility of a world where long-term attachments or even love between robots and humans might exist. The nature of our sexual conduct, in relation to the development of AI, was back on the agenda in 2015 by senior research fellow in the Ethics of Robotics, Dr Kathleen Richardson. Richardson launched a campaign which called for the ban on development of robots and other forms of AI for sexual use arguing: 'it would be unethical and will harm humanity' (Griffin, 2015). Richardson saw the extension of the sex industries and prostitution into the realm of AI furthering the stereotyping of women and the detrimental effects this has on human relations.

Technology may be developed for specific purposes but how it subsequently evolves is often the result of shared or common practices. These may raise particular ethical considerations on which institutions, the legal system and the state have yet to form a coherent policy. However, legislation for regulating the kinds of usage and control has historically been piecemeal. There are examples, throughout history, of how technology emerges and through common use is adapted both for the social good of society but also to its detriment. Indeed, identifying the differences between the good, the bad or the justifiable are part of ethics. Both the sex industries and security services (weapons and warfare) are usually among the first to see the potentials in any technological advancements and quick to develop specific usages. These are also two of the industries that provoke extreme and polarised opinions when questions around morality are debated. Gianmarco Veruggio introduced the term *roboethics* and suggested there were moral obligations on how humans design and use robotic technologies and artificial intelligence (Podcast: Talking Robots, 2008).

We can see in some of the case studies how media fiction presents possible and even credible scenarios for how our world might develop in the future and the ethical dilemmas raised by the interactions of humans and machines. How would state and society evolve and what new legislation would emerge to govern and control the human/ machine society? Literature, film and TV have been exploring the current landscape and initiating debates around the governance and regulation of AI and some of the challenges ahead.

## 9.2 Ethics and Documentary

In the previous sections of this chapter we examined some ethical questions surrounding fiction in relation to written text, film and TV drama and serials. In the following two sections we turn to the more contentious areas of non-fiction work and the role ethics plays in how we represent actuality. Documentary occupies a privileged position in how it is viewed and received by an audience as a medium for 'truth-telling'. What defined traditional documentary forms was its emergence as a medium linked to empirical life experiences (Austin and De Jong, 2008). It was rooted in the factual, evidenced-based view of looking at the world and linked to ideas around documentation and the document; in its perceived ability to bear witness to events that have unfolded or are unfolding in the real world and in its indexical links to actuality. Photography and the moving image in the analogue world were dependent on light sensitive film recording images of something that had material presence. Its validity to depict actuality was reinforced in the legal system where photographs and particular types of recorded film were used as evidence. Documentary traditions were underpinned by the notion of the recording of factual events. They were not staged or made up as in works of fiction nor relied on trained actors to perform a role or be a particular character. This approach can sometimes mask the constructed nature of documentary practice so it is worth repeating John Grierson's famous quote: "the creative treatment of actuality" to remind ourselves of the techniques of filming, framing and editing work that goes into producing a 'story'.

There are key differences between the 'document', 'to document' and 'documentary' that emerged as modes of practice. It is also important to highlight the contested nature of traditional notions of documentary practice based on what many (Barrett, 2006; Austin and De Jong, 2008) argue are false assumptions about the transparency of the medium and any one-to-one correspondence with reality. The 'hand-held' shot and the immediacy of mobile phone cameras have become ubiquitous in signifying the 'real'. Filmmaking, including the production of documentaries, autobiography and the video diary or blog, (vlog) is accessible to anyone with a smartphone and an Internet connection and so has blurred the lines between the amateur and professional. Whilst hybrid forms are continuing to emerge there still remains a distinction between work considered to be of factual origin and the genre of fiction and the reception of such work. Scripted reality television programmes present particular difficulties in how they conflate fiction and actuality. Where ideas around truth-value carry such investment for an audience the burden rests on the documentary maker to consider the question of ethics and to negotiate the tensions between telling a story and the expectations of the audience in how the world is represented (see De Vise, 2010; Boyle, 2013: '*BBC fakes animal shots on wildlife documentaries,*' *says top cameraman*). Once the link between documentary and its relationship to actuality is fractured, a loss of credibility inevitably follows. The audience feel deceived and there is a public outcry. Whilst the 'set-ups' may be a normal part of the storyteller's technique, by not revealing them as such, results in what is perceived as dishonesty on part of the filmmakers. Many wildlife films now include a 'behind the scenes' look at the original filming as an add-on section or diaries, which reveal the making of the film and foreground any shots that have included the 'set-ups'; use of drones to replicate bird flight and any studio-based filming.

So whilst the category of documentary may be open to more than one interpretation there are both expectations on the part of the viewer and intentions on the part of the filmmaker that come into play when we consider the construction of work under the documentary label and ethics. Documentary editor and writer Dai Vaughan (1999), in considering the form states:

'What makes a film "documentary" is the way we look at it; and the history of documentary has been the succession of strategies by which filmmakers have tried to make viewers look at the film in this way' (in Austin and De Jong, 2008: 3).

Documentary can be described as a 'negotiated medium' between the intentions of the filmmaker and the expectations of the audiences. There are many modes of documentary (Nichols, 2010) and a thriving hybridisation, which encompass the spectrum from *cinema verite's* observational mode to the more subjective, reflexive and conversational approaches. The mobile-documentary utilising the accessibility and immediacy of smartphone technology is continuing to produce new configurations of the form. In the variety of current practices from the TV documentary to the mobile-doc there are creative ways around constructing work which can foreground technique and still provide an authentic and honest experience for the viewer. The UK wildlife documentaries, illustrated above, that signal to the audience how the programme has been made are one such example. It is part of the responsibilities of the filmmaker to consider where the balance and boundaries lie and taking a critical approach to the methodology of filmmaking within an ethical framework.

Foregrounding the filmmaker's point of view, common in the more subjective modes of documentary, is another device employed in engaging critically with an audience. Michael Moore's *Bowling for Columbine*, (2002, USA) was a direct attack on the gun culture of America with the intention of getting the 'right to bear arms' legislation changed. Moore was unashamedly part of the 'anti-gun' lobby and his film was about radically influencing public opinion. This was made very clear from the outset and gave audiences the opportunity to consider their own beliefs in relation to both the opinions of Moore and the evidence he presented in the film. He also used the common technique of giving those who hold different perspectives a chance to voice their opinions alongside his own.

*The Thin Blue Line* (1988, USA) by Errol Morris uses documentary techniques and dramatic reconstruction to portray the differing accounts of the killing of a Dallas policeman in 1976. This approach positions the audience in the role of evaluating what is the most credible and 'truthful' account. The accused, Randall Adams, was convicted of murder and sentenced to death but was subsequently released when his conviction was overturned. Problematising issues around accounts of the 'truth', representing different viewpoints and foregrounding personal biases of the filmmaker, are part of getting audiences to think critically in terms of contested subject matter. *The Thin Blue Line* is often cited as a reflexive or hybrid documentary with its cinematic approach to reconstructions and "film-noirish beauty" (Williams, 1993: 12). It has also been credited with helping to secure the release of Adams after 12 years in prison for a crime he didn't commit by galvanising public opinion: a form of documentary advocacy. What is perhaps interesting in terms of our discussion around ethics, are the events that occurred after Adams's release and the film's critical acclaim. Filmmaker Errol Morris was sued by Adams over who had exclusive, legal rights to his story. Whilst Adams had signed a contract in prison, he disputed its terms pertaining to both exclusivity and the money he had been paid for the film (Cerone, 1989).

This example raises ethical consideration for the documentary filmmaker relating to the rights of the human subjects of film. Media ethics is concerned with how mediation may affect individuals or communities. Getting informed consent and fully disclosing the intentions or purpose of the film is part of an ethical method of working and the responsibility of the filmmaker. Informed consent is understood to mean that the person giving consent has the legal capacity to do this. That is to say, they are adults (over the age of 18 in the UK) and are of sound mind. Permissions to use children or vulnerable people in film must be granted by parents or legal guardians. Normal practice involves the filmmaker setting out his/her expectations of what is required and how the material is to be used and distributed. Professional filmmakers use industry standard 'consent-forms' which participants sign to confirm they understand and agree to the reasons and conditions for appearing in the film. In higher budget films legal experts are often employed to work on the finer points of law. There are also examples, as in the case of *The Thin Blue Line,* where disagreements occur regardless of contracts; issues become particularly sensitive when ordinary people, that is, non-actors are involved. Other ethical considerations may include clauses regarding confidentiality and what the consequences for the participants involved might be if these are breached.

Exposè documentaries or investigative journalism style pro-grammes can be particularly challenging to human rights and there is a balance to be struck between revealing 'wrong doings' through undercover methods of journalism and protecting the rights of ordinary people. The impact such media may have, in both the short and long term, needs to be considered in detail. For example, what are the implications of filming ordinary people without their consent on mobile devices? What are the consequences of ordinary people filming and putting material on the Internet? Good practice would suggest that in the majority of cases permissions should be sought and people work within the legal framework of the country they are in. Ultimately the responsibility and possible consequences of filming is placed on the individual filmmaker.

In April 2015, Feidin Santana recorded on his mobile phone the fatal shooting of Walter Scott by a North Charleston, police officer in the US state of South Carolina. The footage was used as part of the evidence to suggest that the account given by the police officer was untruthful. Michael Slager, the officer, was later charged with the murder of Walter Scott and was released on bail pending the outcome of a trial. Without the eyewitness footage from Santana's phone it is doubtful whether the conflicting accounts of events surround Scott's death would have come to light along with the questioning of police behaviour and arrest on a murder charge of a police officer. In an interview, Santana talks about his dilemma in capturing the footage and releasing it publically. He wanted to prove the police account was untruthful and that an unarmed man had been shot in the back and yet he knew his life would change the moment he went public with his story, including fear for his own safety (Yuhas, 2015). However, the decision made by Santana to film and release to the public his 'eye-witness' account was based on moral choices. What he personally determined was the 'right thing to do' for the betterment of justice and society. The personal and moral dilemmas of Santana also need to be viewed in a wider cultural context. Slager is a white police officer accused of shooting an unarmed black man. This narrative is being played out in the landscape of race relations in the United States and follows previous high profile incidents involving police and African Americans, race riots and the civil rights movement.

## 9.3  Ethics and Photography

In considering the varied practices that make up photography there are numerous examples that have highlighted ethical issues in the making and distribution of images. War photography is one contentious example. The claims made for photographs depicting events of war, including victims and the military, are usually about witnessing what is unfolding or telling human-interest stories. The image by Nick Ut, Associated Press photographer taken during the Vietnam war of children fleeing a napalm attack in 1972 was amongst a number of iconic images credited with helping to change public opinion in the US from support for the war to a more an anti-war sentiment (see also Griffiths, 1972; *Vietnam Inc*; Don McCullin, 1973). It could be argued, therefore, that this kind of photographic practice serves a moral function in wanting to change the world for the better. The rest of the world has a 'right to know' what is happening, what atrocities are being committed around the globe. By changing attitudes, so the argument goes, positive changes in society can occur. The immediacy of the image, linked to a particular event at particular moments in time, sets the context.

There are of course bigger moral debates to be had about the nature of war in general and the specifics of individual conflicts. Social historians document and present evidence of conflicts across time, which are articulated, debated and critiqued. It is with the passage of time and through numerous and varied accounts, that the complexities of war narratives are unpicked and conclusions drawn. Photography has played a role in providing documented 'evidence' and in constructing and supporting some of those narratives. Let us consider Ut's iconic image of war, the context of its production and its subsequent reproduction and circulation and re-contextualisation. Every use such an image is put to brings with it a fresh set of ethical challenges. In the case of Ut's photograph depicting 9-year-old Kim Phuc, one of the most common questions asked is how journalists and war photographers continue to work in such situations? At what point do they stop doing their job and intervene? Not only does this become a moral choice it is also an individual choice. Press agencies often instruct employees to work without intervening so that they remain objective observers/recorders. Sometimes this becomes impossible particularly when people are dying in front of you. Ut's response was that he couldn't continue working and helped to get the children to a hospital. Whether to intervene on humanitarian ground or not becomes a moral dilemma for war correspondents who may be given the task of trying in impossible situations to remain neutral. However, it is not often that we hear the back-stories or the personal accounts of the subjects and the photographer. What remains is the image frozen in time.

Ut's image was an exception and the focus for a number of exhibitions re-telling the account of how it had been taken. One such exhibition reunited the adult Kim Phuc with Ut. (Bourne, 2015). Phuc expressed mixed feelings about the image as she felt it had defined the course of the rest of her life. 'In the beginning, says Phuc, she hated the photo it embarrassed her. And she struggled with the publicity that surrounded it ... I realized that if I couldn't escape that picture, I wanted to go back to work with that picture for peace. And that is my choice' (CNN; Newton and Patterson, 2015). Phuc set up the Kim Foundation International to help child victims of war so you might argue here there was a positive outcome to how her image was used.

A more current image, which raises similar ethical questions, is of three years old Aylan Kurdi washed up on the beach at Bodrum, Turkey. It was thought responsible for a shift to more humane reporting of the 2015 Syrian migrant crisis in Europe and that anti-migrant rhetoric was toned down after the image was transmitted on news and internet sites across the world. For the photographers who took pictures of the child and those who distributed them, in both official news sites on TV and across the Internet along with people who circulated the image on social media, ethical choices cannot be so easily separated from actions. Whilst the circulation of the image may have had the desired effect in influencing opinion, what we have no way of knowing is the long-term effects of such an image both at the personal — for the family — and political level. In a world saturated with photographs most eventually fade from memory when they are no longer newsworthy. The Internet has changed our relationship to imagery making such photography accessible and retrievable at any point in the future. Predicting the possible effects of such shifts can only be a prediction about the unknown.

Both these images, taken decades apart, come from a social documentary tradition that claims photography has the ability to change social conditions. The BBC chose to release pictures which didn't show Aylan Kurdi's face; other news outlets carried more graphic images with the latter justifying their actions as 'part of the reality of an ongoing crisis' (http://www.bbc.co.uk/news/world-europe-34133210, accessed October 2015). Individuals, or in the case of the deceased, their families, have little in the way of rights to control the way they are depicted at times of war, crisis or trauma. Control and power rest in the hands of the photographer and any agencies they work for.

There are regulatory bodies that set out codes of practice and guidance for the use of imagery and what can or cannot be reported in the news and by the press. The example above, however, shows that even between established industry practices there are variations in what is deemed ethical to publish. The internet has also made it possible to access material not publishable in the home country. IPSO, the Independent Press Standards Organisation in the UK launched in 2014, is the independent regulator of the newspaper and magazine industry charged with upholding specific codes of conduct. It was set up following recommendations by Lord Justice Leveson's inquiry into allegations of phone hacking by some employees of the Murdock owned News International. The IPSO was to replace The Press Complaints Commission with a more robust set of ethical codes of conduct (http://webarchive.nationalarchives.gov.uk/20140122145147/http:/www.levesoninquiry.org.uk/, accessed October 2015).

All members of the press are expected to maintain certain professional standards. The IPSO Editors Code of Practice 'sets the benchmark for those ethical standards, protecting both the rights of the individual and the public's right to know' (https://www.ipso.co.uk/IPSO/cop.html, accessed October 2015).

It is the interpretation of the latter, the public's 'right to know', balanced against personal freedoms and rights of privacy of the individual that are most contentious. The fact is that questions around ethics are often complex and there are no clear definitives around particular courses of action. Practitioners and professionals may be guided by codes set up by professional bodies that serve to regulate practice, however, the News International scandal (O'Carroll, 2014) exposed behaviour that breached not only public and professional ideas of good practice but was, in some cases, also illegal.

So, working within particular laws and codes of practice set by regulatory bodies could be one way of keeping work legal but it may not cover every situation or moral predicament. Ultimately, we return to the principles and moral beliefs of the individual practitioner at any given time and the choices they make as they frame their image, press the shutter and choose, or not, to circulate that image. We can frame this in even bigger debates on social documentary as a practice and the morality in taking and circulating images of the poor, victims of war, crime, trauma, despair *et al*. Whether such images do indeed lead to social justice and in producing a better world is a highly contentious issue. To complicate matters still further, many of these images become commoditised through time and are re-circulated in the form of books and gallery exhibits. Ritchin (2013) describes the frustration of a European photographer that as his fame increased, he found that magazines wanted to show his work not for the social content and issues he was trying to raise but for his authorship: he had become a name and thus his work was more marketable.

Martha Rosler wrote a seminal paper *In, Around, and Afterthoughts: On Documentary Photography* in which she posed questions around the effectiveness of social documentary claiming it testified more to: 'the bravery (or dare we name it?) manipulativeness of the photographer, who entered a situation of physical danger, social restrictedness, human decay or combination of these and saved us the trouble' (1992: 308). Rosler, like Sontag in the 1970s and other commentators on photography, sees the absorption of documentary into mainstream culture and the elevation of the author neutering its ability to disturb, shock and transform. If the kinds of imagery depicted in war photography, news stories and social documentary is ineffective in moving people to action and in helping change social conditions, if it becomes more about selling coffee-table books for entertainment and producing commodities for the art market rather than changing attitudes then, we might ask, how ethical is it to continue producing such images?

Rosler's critique of the more traditional modes of documentary practice is shaped by a pre-digital era. However, her concluding paragraph hints at a new kind of radical documentary, not financially profitable, that exposes abuses, inequalities, racism, sexism, employment conditions, class oppression and so on. This was an area of practice the late Jo Spence's work covered (see *Cultural Sniping: The Art of Transgression*, 1995), so is not absent from 20th century practice. We might think of this kind of photography in the 21st century as a move to photographic advocacy, a photography of empowerment rather than one that falls strictly within a documentary/social documentary label. We might also consider how digitisation has reinforced some of the critiques of the image and its status to represent an accurate depiction of reality.

When the world is saturated by imagery of every kind how can a space be made for pictures that reveal something new about the world? How do we work ethically when the lines are blurred between professional and amateur practice or when such considerations may have become irrelevant to an online community? A more optimistic reading in the development of photographic practice has been to embrace new technology with hybrid and convergent forms emerging as tools that can be utilised in a more transparent and critical way of working. Images are no longer validated as an unquestionable window on the 'truth' but part of a range of visual practices that range from MRI scans to fictional visual narratives and which can help to interpret, inform, question, entertain and challenge us. Social media and the camera phone have produced a more participatory type of photography. Whilst this may have, in reality, produced large numbers of cute kitten photographs online we also have images that can jolt a nations' conscience, such as the drowned Aylan Kurdi. Global audiences can become active participants. Amateur photographers around the world, using mobile phones have been able to bring a unique perspective to news stories not just with the immediacy of the technology but from their personal, subjective stories which may contrast and conflict with the more 'official' narratives. Seeing the world from an alternative viewpoint can broaden our cultural perspective and change opinion.

Ritchin (2013) speaks about these ideas when he examines the concepts of the *useful photographer* and *making pictures matter.* He examines how the digital environment could provide new synergies for effective storytelling and eye-witness accounts. There are also challenges and Ritchin adds a note of caution. As photography becomes part of a participatory online culture and in blurring boundaries between the photographer and the reader, he asks: How are people protected? I would add that considering how we work and working ethically forces us to engage in a range of strategies and a methodology that demands we consider the impact the images we produce have on others in both the short and long terms.

## 9.4 Ethics, Representation and Human Rights

In considering documentary practice Rabiger comments: 'all stories include assumptions about the way things are. If your film unconsciously reinforces questionable norms, then your embedded values are in the driving seat' (2009: 325). We could apply this not only to all media that we produce but also extend it to ways of working. Rabiger is getting us to consider the ways that certain beliefs become normalised and embedded, often unconsciously, into our ways of seeing the world and everyday practices. In previous sections to this chapter, we have come across such practices when we consider how representations of particular ethnicities or genders are depicted in popular TV programmes or films or even missing completely from our cultural landscape. If there are fewer roles for woman in film and TV than men and they are narrowed further into stereotypes based on age, colour of skin and body shape, then it says something about how women in society are valued. If black men are repeatedly depicted as either drug dealers or rap artists from the myriad of roles they could play then not only does it reinforce and normalise these images within culture but also says something about how black actors within the media industries are valued.

Many countries have started to acknowledge there has been and still remains forms of discrimination and some attempts have been made to rectify this situation through legislation. In the UK, for example, the Equality Act of 2010 (Gov. UK., 2015) is concerned with employability, which makes it illegal to discriminate at work, in education, as consumers, using public services on the grounds of age, race, religion, sex or sexual orientation or disability. These are known as 'protected characteristics' and follow articles outlined in the United Nation's Universal Declaration of Human Rights.

The Universal Declaration of Human Rights was adopted by the United Nations in 1948 in the aftermath of the Second World War. Its 30 articles outline basic human rights and have been the foundation for Human Rights law and international treaties to protect citizens. Many countries have enshrined some of these articles within their own legal systems, such as the anti-discrimination laws, while other articles remain an ideal to aspire to. We can also think about nations where some human rights are not considered a priority and there are examples 'in the interests of national security' where individuals may lose such freedoms enshrined in human rights law. Treaties can be broken or the laws and practices of individual countries may conflict with a more universal understanding of human rights. The difficulty in drawing up universal rights is not simply a case of coming to an agreement on the articles but a matter of how they are interpreted.

For example, Article 19 states 'Everyone has a right to freedom of opinion and expression; this right includes freedoms to hold opinions without interference and to seek, receive and impart information and ideas through any media and regardless of frontiers' (http://www.un.org/en/documents/udhr/index.shtml, accessed October 2015). We can see how this forms the foundation that many countries, including the UK, support to protect freedom of speech. However, this is not without controversy. Some of the views some people may wish to express might be considered extreme or at odds with the laws of the relevant country. In a democracy part of our democratic right is to be heard but it also means others have the right to hold different opinions and to challenge us on ours. It doesn't negate the other human rights, which includes Article 1: 'All human beings are born free and equal in dignity and rights. They are endowed with reason and conscience and should act towards one another in a spirit of brotherhood' (*ibid.*).

It would be interesting to pause for a moment and consider this statement through the lens of gender politics; draw attention to the language used in the 1948 declaration and pose a question we have asked before. We might consider the unconscious biases of language used that may discriminate, albeit, subtly. What might be a more inclusive term than 'brotherhood' that doesn't exclude any members of the population *and* carries with it the sentiment of Article 1 that uphold the ideas of equality for all? You may think I'm labouring a minor point here when the meaning of Article 1 is perfectly clear but I make no apologies for pointing out in this small example how inequalities may be embedded in the terminology and language we use, and which, over time, become normalised. Language constructs meaning and how we are represented in and through language reflects values and beliefs. In an Article as significant as this, which is about how we value each other and act towards each other, we need to be mindful of not excluding — by way of the language we employ — sections of humanity. Consider how this could work for your own projects. Are there less gendered, appropriate terms you might choose over others that are clearly gendered? Some words once taking a masculine form such as *actor* have evolved to be more gender neutral in the 21st century and represent both male and female roles. The diminutive *actress* is now used infrequently in the UK and the USA.

To work ethically, therefore, is to show consideration to those we work with. It forces us to think about our responsibilities to the subjects of our media projects. It asks us to consider methodologies that are fair and largely transparent and ensures we are not setting out to consciously deceive. It also requires of us to think further about our own knowledge and beliefs, where ideas originate from and if we might be unconsciously seeing the world in a particular way, what are the possibilities for looking at something differently?

## 9.5  Ethics and the Internet

When we are considering questions of ethics we can start by asking what are our moral responsibilities as to what we put on screen, record in sound and send out to an audience? How do we represent people in a particular way? How might we use and acknowledge other people's ideas in shaping our own projects? The Internet has been heralded as a new public sphere and that the ease and freedoms of obtaining and sharing information and ideas without restraint is both liberating and part of our democratic rights to be heard. As such it is seen as a force for liberty, freedom and democracy. It has been an enabling tool for artists and filmmakers allowing access to global distribution networks that have, hitherto, been part of a highly industry-controlled field. Work can be created, shared and critiqued freely and can be inspirational or have the power to shock us profoundly. This democratisation means anyone can add content to the web and as a consequence many things found online may not meet the standards or ethical rigour of professionally made and distributed work that is regulated by the media industries. This is especially relevant to practices that are validated for their authenticity and accuracy in reference to actual events. It also offers the possibility for strange juxtapositions and montages which, devoid of any connecting narrative, means Jihadi militant beheading of prisoners are just a click away from the more prosaic blogs of family, pets and beauty products (Ritchin, 2013). There is a platform for everything and the dark web in particular, is testament to the darker aspect of humanity with a thriving community operating in, what has been described with analogies, to the *Wild West*; a kind of Frontier land without boundaries, or the rule of law. How we negotiate this is largely still up to the individual not withstanding the imposition of search engine hierarchies determining some of our browsing choices for us.

We have examined how historical specificities can influence both subject matter and form at any particular moment and also how work can be re-contextualised so meanings shift and slip and may bear no resemblance to the original context of production. We have examined the shifting set of practices that have defined both film and photography and some methodologies that raise ethical concerns about how the world and its inhabitants are represented. Analogue photography wasn't free of manipulation with composition, editing and darkroom techniques contributing to the various subjectivities of the final print. Digitisation raised questions about the nature of authenticity versus simulacra. The idea of the 'faked' image was at the forefront of debates on early digital culture but seems not to have produced the great ethical dilemmas predicted. Bate suggests, whilst digitisation may produce a more critical audience there has been little loss for the veracity and truth-value of everyday news images. They are still largely taken at 'face-value' (2009).

However, digitisation does offer the potential for people to alter or add to existing work in a way that was not previously possible without the author's knowledge and this is more likely to be of issue in art practice. *Appropriation Art* is not a new phenomenon having evolved as a specific genre of art practice before the arrival of the Internet, raising with it questions of copyright and ownership. The web's ability to amplify existing practice and escalate new forms of practice, once again brings to the fore debates around copyright and plagiarism. Copyright law (see https://www.gov.uk/topic/intellectual-property/copyright, accessed October 2015) offers some protection to the originators and authors of creative material to control the way their work is used and distributed but it can be difficult to enforce across the web. Appropriation Art can be something of a grey area and the appropriator usually has to demonstrate that a substantial *new* work has been created from the original source material. Copyright law recognises parody and pastiche, much utilised in online *virals* and popular as a marketing device. This kind of material only succeeds in conveying meaning if the viewer is in on the joke. That is, we are familiar with the original material and recognise what the new work is alluding to while, at the same time, seeing the skill and creativity involved.

There have, however, been a number of high profile cases warranting media scrutiny and which have raised plagiarism and copyright discussions. One such case was the exhibition by Richard Prince, entitled

*New Portraits* (2014), first shown at the Gagosian Gallery, New York and then in London, 2015 (http://www.gagosian.com/exhibitions/richard-prince--september-19-2014, accessed October 2015). The *new* portraits consisted entirely of images Prince obtained from the Instagram website to which he added his own annotations. He went on to sell some of the images for a reported £63,700 (Parkinson, 2015). Some of the people depicted in the photographs were unhappy their images had been appropriated in such a way and that Prince hadn't sought their permission. It illustrates, however, the complexity in applying copyright law in terms of using work taken from social media sites where the *raison d'être* is sharing images with others.

Websites have a legal requirement to remove anything where it can be proved copyright has been breached and prosecution for infringements do occur. Working ethically, therefore, means giving respect and due consideration to other artists and producers and recognising the labour that goes into making media. Most would agree that directly copying artwork or music and trying to pass it off as one's own is morally wrong. Sampling, mash-ups, remixes and parodies can, however, enrich a culture. It forces the audience to consider the original artist and the work and engages with the new work on different terms. It is important, particularly in student work, to document and *cite* the original source of inspiration. This recognises that artists have *intellectual property rights*. Students working within academic institutions such as colleges or universities will find there is usually a clear policy on ethics along with guidance on referencing and ways to avoid plagiarism.

The Internet exposes us to immediacy and variety. Disparate content is juxtaposed though context may be missing. The viewer/user is left to negotiate the possible narratives and meanings as they click through the threads, links and searches. For many this can be both an exciting and frightening prospect. At the same time, we might consider what the web offers producers and distributors in the way of bringing global communities together. How, with the challenges this would inevitably present, sharing practice may offer new possibilities, opportunities and perspectives.

## 9.6 Conclusion

This chapter has been concerned with what it might mean to consider the ethical responsibilities we have when producing work about and with, others. The study of ethics is complex and what shapes our own moral compass is determined by numerous factors including our social and cultural landscape. Moral principles are not a fixed set of beliefs and we have investigated how ideas are shaped across history, time and space. The case studies on TV, documentary and photography have illustrated how morality becomes embedded into culture and how the more nuanced work of the 21st century has repositioned us in a world without clear absolutes on what is right or wrong. We are living through an age of rapid change where digital technology and Web 2.0 continue to have a profound effect on how many of us live our day-to-day lives and how we connect to others. Whilst people remain at the core of our world, developments in quantum computers, artificial intelligence, genetics and robotics point to a future where this may not always be so. The uncertainty of how we might evolve and share our lives with technology is shaping current ethical debates.

There are also continuities with a previous non-digital age when considering how we might think about others. The emergence of the Declaration of Universal Rights after the Second World War was designed to provide a road map for a more civilised world and the agreement of treaties between nations acknowledged basic human rights and the foundation of Human Rights law. We can point to many situations where the law is ambiguous, personal rights are violated or where human rights are still non-existent and yet, throughout the world, a will seems to exist in recognising everyone should have some fundamental rights.

The Internet, if nothing else, has provided a platform for those who have access (there are still areas of the world that don't) for sharing and being part of online communities, bound together by common interests irrespective of geographical constraints or cultural difference. Finding that we have things in common with others is a beginning for mutual understanding and empathy. Homi Bhabha noted empathy creates the condition for human rights and that it is essential for the acknowledgement of others (Bhabha in Anderson, 2009). Questions of ethics can be brought into every aspect of our lives both at work and in the home and part of what makes us human extends to considering how we treat others. This means that at some time in the future ethics in relation to 'others' may extend to acknowledging non-human, artificial intelligent beings and the need for a more inclusive universal rights legislation that embraces a hybrid human-machine age. In the field of media production one of the purposes of including a chapter on ethics is to draw attention to the idea of respecting the rights of others when we go about our research, production and dissemination of media projects. Not everybody has someone to protect their interests. We have also seen that ideas around behaviour and morality are not simply to do with considering only what is deemed right or wrong. Bhabba's ideas around *cultural hybridity* may be more relevant than ever as we embrace an increasingly digitised, technological future. A dynamic, plural society can, through cultural hybridity, offer a "third space" where the "multi lineages of culture" are exposed and offer up creative potential (Johnston and Richardson, 2012). This both recognises others, and allows space for other positions to emerge.

# Bibliography

Abbot, K. (2013) *Tracey Emin and Sarah Lucas: How We Made the Shop.* Available at: http://www.theguardian.com/artanddesign/2013/aug/12/tracey-emin-sarah-lucas-shop (accessed April 2016).

Accessible Arts. Available at: http://www.aarts.net.au/resources/ (accessed June 2015).

Ackroyd, P. (2000) *London: The Biography.* London: Chatto & Windus.

Adam, B. (2003) Reflexive Modernization Temporalized in *Theory, Culture and Society,* 20(2), 59–78.

Adams, R. and Savran, D. (2002) *The Masculinities Studies Reader.* Malden, MA & Oxford: Blackwell Publishers Ltd.

Addley, E. (2014) Ellen's Oscar selfie most retweeted ever- and more of us are taking them. Available at: http://www.theguardian.com/media/2014/mar/07/oscars-selfie-most-retweeted-ever (accessed July 2015).

Agger, B. (2011) iTime: Labor and Life in a Smartphone Era in *Time & Society,* 20(10), 119–136.

Albury, K. (2015) Selfies, Sexts, and Sneaky Hats: Young People's Understandings of Gendered Practices of Self-Representation in *The International Journal of Communication* 9, 1734–1745.

Allan, S. (2007). Citizen Journalism and the Rise of "Mass Self-Communication": Reporting the London Bombings in *Australian Media Journal,* 1(1), 1–20.

The Alzheimer's Society. Available at: http://alzheimers.org.uk (accessed February 2015).

Anderson, A. (2009) No such thing as good and evil: The wire and the humanization of the object of risk in the age of biopolitics. Available at: http://www.darkmatter101.org/site/2009/05/29/no-such-thing-as-good-and-evil-the-wire-and-the-humanization-of-the-object-of-risk-in-the-age-of-biopolitics/ (accessed September 2015).

Anderson, B. (1983) *Imagined Communities: Reflections on the Origin and Spread of Nationalism.* London: Verso.

Anthony, A. (2013) Mandela funeral selfie adds to image problems for Denmark's prime minister. Available at: http://www.theguardian.com/world/2013/dec/14/helle-thorning-schmidt-selfie-mandela-denmark (accessed July 2015).

Appadurai, A. (1996) *Modernity at Large.* Minneapolis: University of Minnesota Press.

Apple Pay print advertisement (2016) *Metro*, April 20, p. 6.

Apple Watch for John Lewis (print advertisement) (2015) *Evening Standard*, October 1, p. 25.

Aspect Ratio. Available at: http://calculateaspectratio.com/photo-video-aspect-ratio (accessed April 2016).

Audioboom. *Sounds of our Shores.* Available at: https://audioboom.com/channel/soundsofourshores (accessed June 2015).

Augé, M. (2004) *Oblivion.* Minneapolis: University of Minnesota Press.

Augé, M. (2008) *Non-Places: An Introduction to Supermodernity.* London: Verso.

Austin, T. and De Jong, W. (2008, ed.) *Rethinking Documentary New Perspectives, New Practices.* Maidenhead and New York: McGraw Hill.

Baker, C., Schleser, M. and Molga, K. (2009) Aesthetics of Mobile Media Art in *Journal of Media Practice*, 10(2–3), 101–122.

Bamberg, M., Krug, K. and Ketchum, G. (2011) *Killer Photos with Your iPhone.* Boston, MA: Cengae Learning.

Banks, J. A. and Deuze, M. (2009) Co-creative Labour in *International Journal of Cultural Studies,* 12(5), 419–431. Available at: http://eprints.qut.edu.au/31340/2/c31340.pdf (accessed January 2016).

Barrett, T. (2006) *Criticizing Photographs* (4th ed.). New York: McGraw Hill.

Barringer, T., Devaney, E., Drabble, M., Gayford, M., Livingstone, M. and Salomon, X. F. (2012) *David Hockney: A Bigger Picture.* London: RA.

Barthes, R. (1972) *Mythologies.* London: Paladin.

Barthes, R. (1977) *Image, Music, Text.* London: Fontana Press.

Bate, D. (2009) *Photography: The Key Concepts.* Oxford & New York: Berg.

Baudelaire, C. (1980) The Modern Public and Photography in Trachtenberg, A. (ed.) *Classic Essays on Photography.* Sedgwick: Leete's Island Books.

Bibliography 227

Baudelaire, C. (2012, Kindle ed.). *The Painter of Modern Life*. London: Penguin Books.
Bauman, Z. (1996) From Pilgrim to Tourist — or a Short History of Identity in Hall, S. and du Gay, P. (eds.) *Questions of Cultural Identity*. London: Sage.
BBC Academy. Available at: http://www.bbc.co.uk/academy (accessed September 2015).
BBC News. National Trust asks public to record seaside sounds. Available at: http://www.bbc.co.uk/news/uk-33217018 (accessed June 2015).
BBC News. Sexting boy's naked selfie recorded as a crime by police. Available at: http://www.bbc.co.uk/news/uk-34136388 (accessed September 2015a).
BBC News. Horizon: How video games can change your brain. Available at: http://www.bbc.co.uk/news/technology-34255492 (accessed September 2015b).
BBC News. Intelligent machines: Call for a ban on robots designed as sex toys. Available at: http://www.bbc.co.uk/news/technology-34118482 (accessed October 2015).
Beattie, J. (2013) David Cameron poses for selfie with Barack Obama at Nelson Mandela's funeral. Available at: http://www.mirror.co.uk/news/world-news/nelson-mandela-memorial-david-cameron-2912167 (accessed July 2015).
Beck, U. (1992) *Risk Society: Towards a New Modernity*. London: Sage.
Beck, U. (1994) The Reinvention of Politics: Towards a Theory of Reflexive Modernization in Beck, U., Giddens, A. and Lash, S. (eds.) *Reflexive Modernization: Politics, Tradition and Aesthetics in the Modern Social Order*. Cambridge: Polity Press, pp. 1–55.
Benkler, Y. (2006) *The Wealth of Networks: How Social Production Transforms Markets and Freedom*. New Haven, Conn.: London, Yale University Press.
Bennett, T. (1999) The Exhibitionary Complex in Boswell, D. and Evans, J. (eds.) *Representing the Nation*. London and New York: Routledge: Open University.
Benjamin, W. (2002) (translated by Eiland, H. and McLaughlin, K.) *The Arcades Project*. Cambridge, MA and London: Harvard University Press.
Berman, M. (2010, ed.) *All That's Solid Melts into Air: The Experience of Modernity*. New York and London: Verso.
Beugge, C. (2014) Blogs that make the most money and how to set up your own. Available at: http://www.telegraph.co.uk/finance/personalfinance/

money-saving-tips/10865063/Blogs-that-make-the-most-money-and-how-to-set-up-your-own.html (accessed March 2016).

Bhabha, H. K. (1990) *Nation and Narration*. London: Routledge.

Bhabha, H. K. (2011) Cultural policies as catalyst of creativity in echoing voices: cultural diversity, a path to sustainable development. Available at: http://www.unesco.org/culture/aic/echoingvoices/downloads/echoing-voices.pdf (accessed October 2015).

Blippar.com. Available at: https://blippar.com/en/ (accessed May 2016).

Boczkowski, P. J. (2004) *Digitizing the News*. Cambridge, MA; London: The MIT Press.

Boswell, D. and Evans, J. (1999) *Representing the Nation*. London and New York: Routledge: Open University.

Boyd, A., Stewart, P. and Alexander, R. (2008) *Broadcast Journalism: Techniques of Radio and Television News*. Amsterdam & Oxford: Focal Press.

Bourdieu, P. (1984) *Distinction: A Social Critique of the Judgement of Taste*. Oxen: Routledge Classics.

Bourne, D. (2015) Napalm victim from powerful Vietnam War picture is reunited with photographer at special Manchester exhibition. Available at: http://www.manchestereveningnews.co.uk/news/greater-manchester-news/napalm-victim-vietnam-war-picture-9493928 (accessed October 2015).

Boyle, S. (2013) BBC fakes animal shots on wildlife documentaries, *says top cameraman*. Available at: http://www.standard.co.uk/news/uk/bbc-fakes-animal-shots-on-wildlife-documentaries-says-top-cameraman-8868376.htm (accessed September 2015).

Bradshaw, P. and Rohuma, L. (2011, ed. and 2014, ed.) *The Online Journalism Handbook: Skills to Survive and Thrive in the Digital Age*. Oxon: Routledge.

British Board of Film Classification. Available at: http://www.bbfc.co.uk/what-classification/guidelines (accessed April 2016).

British Library. *Sounds of Our Shores*. Available at: http://www.bl.uk/sounds-of-our-shores (accessed August 2015).

Brooks, P. (1984) *Reading for the Plot: Design and Intention in Narrative*. Cambridge, MA: Harvard University Press.

Brown, M. (2015) Simon Scharma announces new portrait gallery display with attack on selfies. Available at: http://www.theguardian.com/artanddesign/2015/apr/01/simon-schama-announces-new-portrait-gallery-displays-with-attack-on-selfies?CMP=share_btn_link (accessed April 2015).

Buckley, J. (2015) *The River is the River*. London: Sort of Books.

Buckingham, D. (2009) Speaking Back? In Search of the Citizen Journalist in Buckingham, D. and Willett, R. (eds.) *Video Cultures: Media Technology and Everyday Creativity*. Basingstoke: Palgrave Macmillan, pp. 93–114.

Bull, M. (2013). From the iPod to the Smartphone: Navigating the Spaces of the City in *Explorations in Space and Society*, 28, 13–16.

Burgin, V. (2004) *The Remembered Film*. London: Reaktion Books.

Burns, A. (2015) Self(ie) — Discipline: Social Regulation as Enacted through the Discussion of Photographic Practice in *The International Journal of Communication* 9, 1716–1733.

Carmichael Aitchison, C. (2007, ed.) *Sports and Gender Identities*. New York & London: Routledge.

Carson, J. and Kelly, C. (1993–1995) Evening echoes. Available at: http://artscool.cfa.cmu.edu/~carson/evening-echoes.html (accessed March 2016).

Carter, A. (1979) *The Bloody Chamber*. London: Penguin Books.

Cashmore, E. and Cleland, J. (2012) Fans, Homophobia and Masculinities in Association Football: Evidence of a More Inclusive Environment in *The British Journal of Sociology* 63(2) 370–387.

Cellan-Jones, R. (2014) Stephen Hawking warns artificial intelligence could end mankind. Available at: http://www.bbc.co.uk/news/technology-30290540 (accessed September 2015).

Cerone, D. (1989) 'Thin Blue line in a Different Court'. Available at: http://articles.latimes.com/1989-07-07/entertainment/ca-3332_1_thin-blue-line (accessed October 2015).

Cheesman, C. (2015) European Parliament rejects 'absurd' EU plan to axe Freedom of Panorama *Amateur Photographer*. Available at: http://www.amateurphotographer.co.uk/latest/photo-news/european-parliament-rejects-absurd-eu-plan-to-axe-freedom-of-panorama-55708 (accessed July 2015).

Chiang, K. J., Chu, H. and Chang, H. J. *et al.* (2010) The Effects of Reminiscence Therapy on Psychological Well-Being, Depression, and Loneliness Among the Institutionalised Aged in *International Journal of Geriatric Psychiatry*, 25, pp380–388. Available at: http://onlinelibrary.wiley.com/doi/10.1002/gps.2350/epdf (accessed November 2015).

Cook, I. (Arsenal FC historian). Available at: http://www.arsenal.com/history/ (accessed July 2015).

Copyleft. Available at: https://copyleft.org/ (accessed June 2016).

Counter-Terrorism Act. Available at: http://services.parliament.uk/bills/2007-08/counterterrorism/documents.html (accessed June, 2016). Section 76. Available at: http://www.legislation.gov.uk/ukpga/2008/28/section/76 (accessed July 2012 and 2015).

Coverley, M. (2010) *Psychogeography*. Harpenden: Pocket Essentials.

Creative Commons. Available at: https://creativecommons.org/ (accessed May 2016).

CNN Iconic Photos of the Vietnam War. Available at: http://edition.cnn.com/2014/06/19/world/gallery/iconic-vietnam-war-photos/ (accessed October 2015).

Cyber Bullying and Sexting on Social Media. Available at: http://www.ncpc.org/programs/living-safer-being-smarter/surfing-safer/cyberbullying-and-sexting-on-social-media (accessed January 2016).

Dannenbaum, J., Hodge, C. and Mayer, D. (2003) *Creative Filmaking from the Inside Out: Five keys to the Art of making Inspired Movies and Television*. New York, London: Fireside/Simon and Schuster.

Davies, C. (2015) Chelsea fans accused of Paris Métro racism to fight football banning orders. Available at: http://www.theguardian.com/world/2015/mar/25/chelsea-alleged-racism-paris-metro-fans-fight-football-banning-orders (accessed July 2015).

Davies, G. (2010) *Copyright Law for Artists, Photographers and Designers*. London: A and C Black.

Debord, G. (1956) *The Theory of Dérive* in Xavier, C. (1996) *The Dérive and Other Situationist Writings of the City* Museu d'Art Contemporani de Barcelona.

De Jong, W. Knudsen, E. and Rothwell, J. (2012) *Creative Documentary*. Harlow: Longman.

De Lange, M. (2010) *Moving Circles: Mobile Media and Playful Identities*. Dissertation from Erasmus University, Rotterdam. Available at: http://www.bijt.org/wordpress/wp-content/uploads/2010/11/De_Lange-Moving_Circles_web.pdf (accessed December 2015).

*Deliciously Ella Blog*. Available at: http://deliciouslyella.com/ (accessed March 2016).

Deloitte (2015) *Mobile Consumer 2015: The UK Cut*. Available at: www.deloitte.co.uk/mobileuk (accessed June 1, 2016).

Deuze, M. (2005) What is Journalism? Professional Identity and Ideology of Journalists Reconsidered in *Journalism*, 6(4), 442–464.

Deuze, M. (2007) Convergence Culture in the Creative Industries in *International Journal of Cultural Studies*, 10, (2), 243–263.

De Vise, D. (2010) Wildlife filmmaker Chris Palmer shows that animals are often set up to succeed. Available at: http://www.washingtonpost.com/wp-dyn/content/article/2010/09/21/AR2010092105782.html (accessed September 2015).

Discrimination Your Rights, Government UK (2015). Available at: https://www.gov.uk/discrimination-your-rights/types-of-discrimination (accessed October 2015).

Duffin, D. (1991) *Artists Handbook* 5: *Organising Your Exhibition*. Sunderland: AN Publications.

Elkin, L. (2016) *Flâneuse: Women Walk the City in Paris, New York, Tokyo, Venice. London*. London: Chatto & Windus.

Elliott, A. and Urry, J. (2010) *Mobile Lives*. New York & London: Routledge.

Ellis, K. and Goggin. G. (2015) *Disability and the Media*. London & New York: Palgrave.

Elsey, E. and Kelly, A. (2005, ed.) *In Short: A Guide to Short Filmmaking in the Digital Age*. London: BFI Publishing.

Englander, E. (2014) Stop demonizing teen sexting. In most cases, it's completely harmless. Available at: https://www.washingtonpost.com/posteverything/wp/2014/11/07/stop-demonizing-teen-sexting-in-most-cases-its-completely-harmless/ (accessed August 2015).

Equity. Available at: https://www.equity.org.uk/resource-centre/for-employers/ (accessed May 2016).

Feifer, J. (2013) Selfies at serious place. Available at: http://selfiesatserious-places.tumblr.com/ (accessed July 2015).

Feifer, J. (2015) *Is this a selfie?* Available at: http://www.nytimes.com/2015/07/22/opinion/is-this-a-selfie.html?smid=tw-nytopinion&_r=0 (accessed July 2015).

Fernyhough, C. (2012) *Pieces of Light: The New Science of Memory*. London: Profile Books Limited.

FilmicPro.com. Available at: http://www.filmicpro.com/apps/filmic-pro/ (accessed May 2016).

Fletcher, J. (2013) Open wide for Chernobyl from selfies at serious places. Available at: http://selfiesatseriousplaces.tumblr.com/ (accessed July 2015).

Florida, R. (2002) *The Rise of The Creative Class*. New York: Basic Books.

Foot, K. A. (2002) Online Action in Campaign 2000: An Exploratory Analysis of the U.S. Political Web Sphere, *Journal of Broadcasting and Electronic Media*. Available at: https://people.sunyit.edu/~steve/foot-schneider-online-action-jbem.pdf (accessed January 2016).

Football Related Arrests and Banning Orders. Available at: https://www.gov.uk/government/uploads/system/uploads/attachment_data/file/352864/FootballRelatedArrestsBanningOrderSeason201314.pdf (accessed July 2015).

Fox, A. and Caruana, N. (2012) *Behind the Image*: *Research in Photography*. Lusanne, London: AVA Publishing.

Frosh, P. (2015) The Gestural Image: The Selfie, Photographic Theory and Kinesthetic Sociability in *International Journal of Communication*, 9(2015), 1607–1628.

Giddens, A. (1991) *Modernity and Self-Identity*: *Self and Society in the Late Modern Age*. Cambridge: Polity Press.

Goggin, G. and Hjorth, L. (2014, ed.) *The Routledge Companion to Mobile Media*. New York & London: Routledge.

Goldblatt, D. (2014) David Goldblatt on the most political matches in history in *Social Science*. Available at: http://www.socialsciencespace.com/2014/06/david-goldblatt-on-the-sociology-of-football/ (accessed July 2015).

Goode, L. (2009) Social News, Citizen Journalism and Democracy in *New Media & Society*, 11(8), 1287–1305.

Greenberg, R., Ferguson, B. W. and Nairne, S. (2009, ed.) *Thinking About Exhibitions*. London & New York: Routledge.

Greenfield, S. (2011) Susan Greenfield: Living Online is Changing our brains, *New Scientist*, July 2011. Available at: https://www.newscientist.com/article/mg21128236-400-susan-greenfield-living-online-is-changing-our-brains/ (accessed August 2015).

Griffin, A. (2015) Sex Robots Should be Banned, Say Campaigners, as Engineers Look to Add AI to Sex Toys. Available at: http://www.independent.co.uk/life-style/gadgets-and-tech/news/sex-robots-should-be-banned-say-campaigners-as-engineers-look-to-add-ai-to-sex-toys-10501622.html (accessed October 2015).

Griffiths, P. J. (2006, ed.) *Vietnam Inc*. London: Phaidon Press.

Gye, L. (2007) Picture This: The Impact of Camera Phones on Personal Photographic Practice Continuum in J*ournal of Media and Cultural Studies*, 21(2), 279–288.

Hall, S. (1980) Encoding/Decoding in Hall, S. Hobson, D., Lowe, A. and Willis, P. (eds.) *Culture, Media, Language*. London and New York: Routledge.

Harvey, D. (1990) *The Condition of Postmodernity*. Cambridge: Blackwell.

Hampton, M. (2010) *The Fourth Estate Ideal in Journalism History*. New York: Routledge.

Hawley, S. (2008) *Aesthetics of the Mobile Video,* Filmobile International Conference, University of Westminster, UK.

Healey, M. (2008) *What is Branding?* Hove: RotoVision SA.

Heaphy, B. (2007) *Late Modernity and Social Change: Reconstructing Social and Personal Life.* London: Routledge.

Heathcote, J. (2009) *Memories are Made of This: Reminiscence Activities for Person-Centred Care.* London: Alzheimer's Society.

Hegarty, J. (2014) *Hegarty on Creativity.* London: Thames and Hudson.

Henig, S. (2005) *Citizens, Participants and Reporters* in Columbia Journalism Review. Available at: http://www.cjr.org/politics/citizens_participants_and_repo.php (accessed January, 2016).

Hess, A. (2015) The Selfie Assemblage in *The International Journal of Communication,* 9, 1629–1646.

Hill, S. and Lashmar, P. (2014) *Online Journalism: The Essential Guide.* London: Sage.

Hirsch, M. (2012a) *Family Frames: Photography, Narrative and Postmemory.* Cambridge MA, London: Harvard University Press.

Hirsch, R. (2012b, ed.) *Light and Lens: Photography in the Digital Age.* Oxford: Focal Press.

Hipstamatic.com. Available at: http://hipstamatic.com/camera/ (accessed January 2016).

Hjorth, L. (2005) Locating Mobility: Practices of co-presence and the persistence of the postal metaphor in SMS/MMS mobile phone customization in Melbourne Fibreculture.org. Available at: http://six.fibreculturejournal.org/fcj-035-locating-mobility-practices-of-co-presence-and-the-persistence-of-the-postal-metaphor-in-sms-mms-mobile-phone-customization-in-melbourne/ (accessed January 2016).

Hollows, J. and Jancovich, M. (1995) *Approaches to Popular Film.* Manchester: Manchester University Press.

Holt, K. and Karlsson, M. (2014) Random Acts of Journalism?: How Citizen Journalists Tell the News in Sweden in *New Media & Society,* 17 (11), 1795–1810.

Hornby, N. (1992) *Fever Pitch.* London: Penguin Books.

Hudson, D. and Zimmermann, P. (2015) *Thinking Through Digital Media.* London: Palgrave MacMillan.

*Humans* Channel 4 TV, UK. Available at: http://www.channel4.com/programmes/humans/on-demand/62381-001 (accessed September 2015).

iPhone Film Festival (IFF). Available at: http://www.iphoneff.com/ (accessed June 2016).

The Institute for Research and Innovation in Social Services (IRISS). Available at: http://www.iriss.org.uk/resources/supporting-those-dementia-reminiscence-therapy-and-life-story-work (accessed November 2015).

Internet Encyclopaedia of Philosophy. Available at: http://www.iep.utm.edu/(accessed September 2015).

Intellectual Property Copyright. Available at: https://www.gov.uk/topic/intellectual-property/copyright (accessed October 2015).

Ishiguro, H. (2015) Symbiotic Human — Robot Interaction Project. Available at: http://www.jst.go.jp/erato/ishiguro/en/index.html (accessed January 2016).

Jabour, B. (2015) Hate selfie sticks, or just young women having fun? Available at: http://www.theguardian.com/commentisfree/2015/apr/17/hate-selfie-sticks-or-just-young-women-having-fun (accessed April 2015).

Jacobi, C. and Kingsley, H. (2016) *Painting With Light*. London: Tate Publishing.

Jenkins, H. (2006) *Convergence Culture: Where Old and New Media Collide*. New York; London: New York University Press.

Jenkins, H., Ford, S. and Green, J. (2013) *Spreadable Media*. New York and London: New York University Press.

Johnston, G. (2015) McGrath the latest victim of Trial by Twitter. Available at: http://thenewdaily.com.au/sport/2015/02/22/mcgrath-latest-victim-trial-twitter/ (accessed July 2015).

Johnston, I. and Richardson, G. (2012) Homi Bhabha and Canadian Curriculum Studies: Beyond the Comfort of the Dialectic in *The Journal of the Canadian Association for Curriculum Studies,* 10(1), 115–137.

Johnson, K. A. and St. John III, J. (2015) Citizen Journalists' views on traditional notions of journalism, story sourcing, and relationship building: The Persistence of legacy norms in an emerging news environment, *Journalism Studies*. Available at: http://dx.doi.org/10.1080/14616 70X.2015.1051094.

Johnson, R. (1986) The Story So Far: And Further Transformations? in Punter, D. (ed.) *Introduction to Contemporary Cultural Studies*. Harlow: Longman.

Jones, J. (2015a) RIP the selfie: When prince Harry calls time on a craze, you know it's well and truly dead. Available at: http://www.theguardian.com/artanddesign/jonathanjonesblog/2015/apr/07/selfie-prince-harry-died-in-2015-selfie-stick (accessed April 2015).

Jones, J. (2015b) The Selfie museum: Why big art galleries should take it seriously. Available at: http://www.theguardian.com/culture/

jonathanjonesblog/2015/mar/30/museum-selfie-sticks-banned-photography (accessed March 2015).

Katz, J. E. and Aakhus, M. (2006, ed.) *Perpetual Contact: Mobile Communication, Private Talk, Public Performance.* Cambridge: Cambridge University Press.

Keegan, S. (2014) Digital technologies are re-shaping our brains: What are the implications for society and the research industry? Available online at: www.emeraldinsight.com/1352-2752.htm (accessed August 2015).

Keightley, E. and Pickering, M. (2006) For the record: Popular Music and Photography as Technologies of Memory in the *European Journal of Cultural Studies,* 9(2), 149–165.

Kickitout.org. Available at: http://www.kickitout.org (accessed July 2015).

King, A. (1997) The Lads: Masculinity and the New Consumption of Football in *Journal of Sociology,* 31(2), 329–346.

Kobre, K. (2008) *Photojournalism: The Professionals' Approach* (6th ed.). Oxen: Focal Press.

Kodak Online. Available at: http://www.kodak.com/ek/us/en/corp/aboutus/heritage/milestones/default.htm (accessed April 2016).

Kovach, B. and Rosenstiel, T. (2003) *The Elements of Journalism.* London: Atlantic Books.

Krell, M. (2012) New Media, New Documentary Forms in De Jong, W., Knudsen, E. and Rothwell, J. *Creative Documentary.* Harlow: Longman.

Kress, G. and Van Leeuwen, T. (2009, ed.) *Reading Images: The Grammar of Visual Design.* London and New York: Routledge.

Kuhn, A. (1995) *Family Secrets: Acts of Memory and Imagination.* London & New York: Verso.

Kurt, S. and Osueke, K. K. (2014) *The Effect of Color on the Moods of College Students.* Thousand Oaks: Sage Publication, pp.1–12.

Landsberg, A. (2003) Prosthetic Memory: The Ethics and Politics of Memory in an Age of Mass Culture in Grainge, P. (ed.) *Memory and Popular Film.* Manchester and New York: Manchester University Press.

Lash, S. and Urry, J. (1994) *Economies of Signs and Space.* London: Sage.

Latinus, M. and Belin, P. (2011) Human Voice Perception in *Current Biology,* 21(4). Available at: www.psy.gla.ac.uk (accessed July 2015).

Lewis, P. (2009) Video reveals G20 police assault on man who died. Available at: http://www.theguardian.com/uk/2009/apr/07/video-g20-police-assault (accessed July 2015).

Levy, D. (2007) *Love and Sex with Robots: The Evolution of Human–Robot Relationships.* New York: HarperCollins.

Lexus NX F Sport (TV advertisement) (2016), July. Available at: https://www.youtube.com/watch?v=ta5XUNmd07k (accessed July 20, 2016).

The Listening Project. Available at: http://www.bbc.co.uk/radio4/features/the-listening-project (accessed April 2015).

Long, P. and Wall, T. (2009) *Media Studies: Texts, Production and Contexts.* Harlow: Pearson Education Limited.

Lorenc, J. Skolnick, L. and Berger, C. (2010) *What is Exhibition Design?* Hove: RotoVision SA.

LUCI Software Live video App. Available at: http://www.luci.eu/ (accessed April 2015).

Luessen, C. (2012) *The Flâneur, Psychogeography and Drift Photography.* Available at: http://arthopper.org/the-flaneur-psychogeography-and-drift-photography/ (accessed February 2016).

Lux Online. Available at: www.luxonline.org.uk (accessed April 2015).

Macdonald, K. (2011) Life in a day: Around the World in 80,000 clips. Available at: http://www.theguardian.com/film/2011/jun/07/life-in-a-day-macdonald (accessed March 2016).

Machill, M., Kohler, S. and Waldhauser, M. (2007) The Use of narrative Structures in Television News in *The European Journal of Communication,* 22(2), 185–205.

Macpherson, L. (2004) Photographers' Rights in the UK. Available at: http://www.thecameraclub.co.uk/UKPhotographersRights.pdf (accessed May 2016).

Mass Observation. Available at: http://www.massobs.org.uk/ (accessed August 2015).

Marincola, P. (2006, ed.) *Questions of Practice: What Makes a Great Exhibition?* Chicago: Philadelphia Exhibition Initiative/Reaktion Books Ltd.

McDonald, H. (2012) Nick Clegg hits out at trail by Twitter over child abuse claims. Available at: http://www.theguardian.com/politics/2012/nov/09/nick-clegg-trial-twitter-abuse (accessed July 2015).

Mcfadden, S. (2014) Teaching the camera to see my Skin. Available at: http://www.buzzfeed.com/syreetamcfadden/teaching-the-camera-to-see-my-skin#.dxR96yaJB (accessed January 2016).

Meyer, T. (2008) *Mobile-Mentary: An Approach.* Filmobile International Conference, University of Westminster, UK.

Miller, D. (2011) *Tales from Facebook.* Cambridge & Oxford: Polity Press.

Mirzoeff, N. (2015) *How to see the World.* London: Pelican Books.

The Mobile Film Festival (France). Available at: http://www.mobilefilmfestival.com/ (accessed May 2016).

Mobile Innovations Network Australasia (MINA). Available at: http://mina.pro/ (accessed May 2016).

Monaco, J. (2000, ed.) *How to Read A Film*. New York and Oxford: Oxford University Press.

Moores, S. (2004) The Doubling of Place: Electronic Media, Time-Space Arrangements and Social Relationships in Couldry, N. and McCarthy A. (eds.) *Media Space*. Abingdon: Routledge.

Moore, S. (2015) Give our kids rights over the digital world. Let them be stupid and foolish. Available at: http://www.theguardian.com/commentisfree/2015/jul/29/give-kids-rights-over-their-digital-world-stupid-and-foolish-suzanne-moore (accessed August 2015).

Moran, C. (2013) Time as social practice in time & society, pp. 1–21. Available at: http://tas.sagepub.com/content/early/2013/04/10/09 61463X13478051.full.pdf+html (accessed February 9, 2016).

Nash, R. (2006) Questions of Practice in Marincola P. (ed.) *Questions of Practice: What Makes a Great Exhibition?* Chicago: Philadelphia Exhibition Initiative/Reaktion Books Ltd.

Naughton, J. (2016) Forget ideology, liberal democracy's newest threats come from technology and bioscience in *The Guardian*. Available at: https://www.theguardian.com/commentisfree/2016/aug/28/ideology-liberal-democracy-technology-bioscience-yuval-harari-artificial-intelligence (accessed August 2016).

Newton, P. and Patterson, T. (2015) CNN: The girl in the picture: Kim Phuc's journey from war to forgiveness. Available at: http://edition.cnn.com/2015/06/22/world/kim-phuc-where-is-she-now/ (accessed October 2015).

Nichols, B. (2010) *Introduction to Documentary* (2nd ed.). Bloomington & Indianapolis: Indiana University Press.

Nowotny, H. (2016) *The Cunning of Uncertainty*. Cambridge: Polity.

O'Carroll, L. (2014) Phone-hacking scandal: Time line. Available at: http://www.theguardian.com/uk-news/2014/jun/24/phone-hacking-scandal-timeline-trial (accessed April 2016).

Odin, R. (2012) Spectator, Film and the Mobile Phone in Christie, I. (ed.) *Audiences: Defining and Researching Screen Entertainment Reception*. Amsterdam: Amsterdam University Press.

Oxford Dictionaries Online. Available at: http://www.oxforddictionaries.com/definition/english/selfie (accessed April 2015).

Oxford University Press Blog. Available at: http://blog.oup.com/2013/11/selfies-history-self-portrait-photography/ (accessed July 2015).

Papacharissi, Z. (2011) *A Networked Self: Identity, Community, and Culture on Social Network Sites*. New York & London: Routledge.

Park, R. E. (1940) News as a Form of Knowledge: A Chapter in the Sociology of knowledge in *American Journal of Sociology*, 45, pp. 669–686. University of Chicago Press Journals online available at: http://www.jstor.org/stable/pdf/2770043.pdf (accessed March, 2016)

Parkinson, H. J. (2015) *Instagram, an artist and the $100,000 selfies — appropriation in the digital age*. Available at: http://www.theguardian.com/technology/2015/jul/18/instagram-artist-richard-prince-selfies (accessed October 2015).

PayPal App (print advertisement) (2016) *Metro*, April 19, p. 11.

Pearsall, J. (2013) Available at: http://blog.oxforddictionaries.com/press-releases/oxford-dictionaries-word-of-the-year-2013/ (accessed July 2015).

Penenberg, A. (2009) *Viral Loop*. New York: Hyperion.

Percy, M. and Taylor, R. (1997) Something for the Weekend, Sir? Leisure, Ecstasy and Identity in Football and Contemporary Religion, *Leisure Studies* 16 (1), 37–49.

Phelan, D. (2015) Apple watch review: First look at Apple's smartwatch ahead of launch in *The Independent*, April 8. Available at: http://www.independent.co.uk/life-style/gadgets-and-tech/news/apple-watch-first-look-review-hands-on-with-apples-highly-anticipated-smart-watch-10161523.html (accessed April 18, 2016).

Plant, S. (2002) On the Mobile: The effects of mobile telephones on social and individual life. A study commissioned by Motorola. Available at: http://www.momentarium.org/experiments/7a10me/sadie_plant.pdf (accessed January 8, 2016).

The Pocket Cinema Festival. Available at: http://pocketcinemaff.com/ (accessed June 2016).

Powell, H. (2012) *Stop the Clocks! Time and Narrative in Cinema*. London: I.B. Tauris.

Powell, H. (2013) *Promotional Culture and Convergence: Markets, Methods, Media*. Oxon & New York: Routledge.

Powell, H. and Prasad, S. (2007) Lifeswap: Celebrity Expert as Lifestyle Advisor in Heller, D. (ed.) *Makeover Television: Realities Remodelled*. New York: I.B. Tauris.

Propp, V. (1927) Scott, L. and Wagner, L. (1969, eds.) *Morphology of the Folk Tale*. Austin: University of Texas Press.

Pullman, P. (2013) *Grimm Tales for Young and Old*. London: Penguin Classics.

Quinn, S. (2005) What is Convergence and How will it Affect my Life? in Quinn, S. and Filak, V. (eds.) *Convergent Journalism: An Introduction*. Waltham: Focal Press, pp. 8–16.

Rabiger, M. (2009) *Directing the Documentary* (5th ed.). New York & London: Focal Press.

Raindance Film Festival. Available at: http://raindancefestival.org/ (accessed June 2016).

Raindance Web Festival. Available at: https://filmfreeway.com/festival/ RaindanceWebFest (accessed June 2016).

Rawlings, K. (2013) Selfies and the History of Self-portrait Photography. Available at: http://blog.oup.com/2013/11/selfies-history-self-portrait-photography/ (accessed April 2016).

Reda, J. (2015) Freedom of Panorama Under Threat. Available at: https:// juliareda.eu/2015/06/fop-under-threat/ (accessed July 2015).

Reeser, T. W. (2010) *Masculinities in Theory*. Oxford: Wiley-Blackwell.

Rehling, N. (2011) *"It's About Belonging"*: *Masculinity, Collectivity, and Community* in British Hooligan Films. *Journal of Popular Film & Television*, 39(4), pp. 162–173.

Renov, M. (1993, ed.) *Theorizing Documentary* London: Routledge.

Rettberg, J. W. (2014) Seeing ourselves through technology. Available at: http:// www.palgraveconnect.com/pc/doifinder/10.1057/9781137476661. 0007 (accessed August 2015).

Ricoeur, P. (1984) *Time and Narrative I*. Chicago: The University of Chicago Press.

Ritchin, F. (2013) *Bending the Frame*: *Photojournalism, Documentary and the Citizen*. New York: Aperture.

Rix Media. Available at: www.rixmedia.org (accessed January 2016).

Robinson, J. (2005) Creating a web based archive of dialect recordings. Available at: http://library.leeds.ac.uk/emumedia/2162/JonathanRobinson.pdf (accessed August 2015).

Rosen, J. (2006). The people formerly known as the audience. PressThink: Ghost of Democracy in the Media Machine, June 27. Available at: http://archive.pressthink.org/2006/06/27/ppl_frmr.html (accessed January 2016).

Rosler, M. (1992) In, Around and Afterthought on Documentary Photography in Bolton, R. (ed.) *The Contest of Meaning*. Cambridge, MA & London: MIT Press.

Rugoff, R. (2006) You Talking to Me? On Curating Group Shows That Give You a Chance to Join the Group in Marincola P. (ed.) *Questions of*

*Practice*: *What Makes a Great Exhibition?* Chicago: Philadelphia Exhibition Initiative/Reaktion Books Ltd.

Rushdie, S. (2015) *Two Years, Eight Months and Twenty-Eight Nights.* London: Jonathan Cape.

Saltz. J. (2014). At Arm's Length: A History of the Selfie. *New York Magazine.* Available at: http://www.vulture.com/2014/01/history-of-the-selfie.html (accessed July 2015).

Sambrook, R. (2005) Citizen Journalism and the BBC, Nieman Reports: Nieman Foundation for Journalism at Harvard. Available at: http://www.encoreleaders.org/wp-content/uploads/2013/06/Nieman-Reports-_-Citizen-Journalism-and-the-BBC.pdf (accessed December 2015).

Schacter, D. (1996) *Searching for Memory: The Brain, the Mind and the Past.* New York: Basic Books.

Schleser, M. (2011) *Mobile-Mentary: Mobile Documentaries in the Mediascape.* Saarbrücken, Germany: LAP Lambert Academic Publishing.

Schleser, M. (2014). Connecting through Mobile Autobiographies: Self-Reflexive Mobile Filmmaking, Self-Representation, and Selfies in Berry, M. and Schleser, M. (eds.) *Mobile Media Making in an Age of Smartphones.* London: Palgrave MacMillan, pp. 148–158.

Schleser, M. R. C. (2013) MINA — Mobile Innovation Network Aotearoa and "FILMOBILE" in *Ubiquity: The Journal of Pervasive Media,* 2(1+2), 105–115.

Schleser, M. R. C. and Turnidge, T. (2013) 24 Frames 24 Hours in *Ubiquity: The Journal of Pervasive Media,* 2(1+2), 205–213.

Schroder, K. (2000) Making Sense of Audience Discourses: Towards a Multidimensional Model of Mass Media Reception in *The European Journal of Cultural Studies,* 3(2), 233–258.

Scott, D. M. (2010) *The New Rules of Marketing and PR.* Hoboken, NJ: John Wiley and Sons.

Seaborne, M. and Sparham, A. (2011) *London Street Photography 1860–2010.* London: Museum of London and Dewi Lewis Publishing.

Self, W. (2016) *Our Digital Lives and the Chaos Beneath.* Available at: https://www.theguardian.com/books/2016/aug/06/will-self-digital-lives-chaos-box-set (accessed August 2016).

Senft, T. and Baym, N. (2015) What Does the Selfie Say? Investigating a Global Phenomena in *The International Journal of Communication,* 9, 1588–1606.

SelfieCity. Available at: http://selfiecity.net/ (accessed July 2015).

Settle, M. (2015) BBC Academy. Available at: http://www.bbc.co.uk/academy/news/article/art20130711164645373 (accessed April 2015).

Sheller, M. (2004) Mobile Publics: Beyond the Network Perspective in *Environment and Planning D: Society and Space, 22*, 39–52.

Simmel, G. (1903/1971) *The Metropolis and Mental Life* in *Georg Simmel: On Individuality and Social Forms*. Edited and with an introduction by D. N. Levine. Chicago: The University of Chicago Press.

Sinclair, I. (1997) *Lights Out For The Territory*. London: Granta Books.

Sinclair, I. (2002) *London Orbital: A Walk Around the M25*. London: Penguin Books.

Sinclair, I. (2009) *Hackney, That Red-Rose Empire*. London: Hamish Hamilton.

Sinembago.mx (2013) Available at: http://www.sinembargo.mx/28-07-2013/700284 (accessed February, 2016).

Sischy, I. (1991) Photography: Good Intentions in *The New Yorker*, September 9, 89–95. Available at: http://archives.newyorker.com/ (accessed September, 2016).

Skeggs, B. (2002, ed.) *Formations of Class and Gender: Becoming Respectable*. London: Sage Publications Ltd.

Soltis, A. (2013) *Michelle not Amused by Obama's Memorial Selfie*. Available at: http://nypost.com/2013/12/10/michelle-annoyed-by-obamas-selfie-at-mandela-memorial/ (accessed July 2015).

Sontag, S. (1977) *On Photography*. London: Penguin Books.

Slater, B. H. (2015) Aesthetics in the Internet Encyclopedia of Philosophy. Available at: http://www.iep.utm.edu/aestheti/ (accessed June 2015).

Smith, M. J. (1998) *Social Science in Question*. London: Sage Publications.

Snapseed. Available at: https://support.google.com/snapseed/?hl=en#topic=6155507 (accessed June 2016).

Spence, J. (1986) *Putting Myself in the Picture*. London: Camden Press.

Spence, J. (1995) *Cultural Sniping: The art of Transgression*. London and New York: Routledge.

Spence, J. and Holland, P. (1991) *Family Snaps: The Meanings of Domestic Photography*. London: Virago Press.

Steensen, S. (2011) Online Journalism and the Promises of New Technology in *Journalism Studies*, 12(3), 311–327.

StoryCorp. Available at: http://storycorps.org/ (accessed April 2015).

Stubbs, G. (2014) World's oldest selfies: Pictures from as far back as 1839 show craze isn't so modern. Available at: http://www.mirror.co.uk/

news/weird-news/worlds-oldest-selfies-pictures-far-3894621 (accessed July 2015).

Subrahmanyam, S. (2015) Real people are censored, the anonymous say what they want. Interview with Sagarika Ghosh. Available at: http://timesofindia.indiatimes.com/home/sunday-times/all-that-matters/Real-people-are-censored-the-anonymous-say-what-they-want-Sanjay-Subrahmanyam/articleshow/50224623.cms (accessed December 2015).

Suggs, Z. (2014) *Girl Online*. London: Penguin Books.

Suggs, Z. Zoella: Beauty, Fashion, Lifestyle Blog. Available at: https://www.zoella.co.uk/ (accessed March 2016).

Sundaram, R. (2015) Post-Postcolonial Infrastructure. *Eflux*. Available at: http://www.e-flux.com/journal/post-postcolonial-sensory-infrastructure/ (accessed December 2015).

Syrian Mobile Films. Available at: http://syriamobilefilms.com/en/ (accessed June 2016).

Tagg, J. (1998) *The Burden of Representation*. Basingstoke: Palgrave Macmillan.

Techradar. Available at: www.techradar.com (accessed June 2015).

Thompson, E. P. ([1967] 1991) *Time, Work-Discipline and Industrial Capitalism* in *Customs in Common*. London: Merlin Press.

Tifentale, A. (2014) The Selfie: Making sense of the "Masturbation of self image" and the "virtual mini-me". Available at: http://selfiecity.net/ (accessed August 2015).

The Toronto Smartphone Film Festival. Available at: http://www.smart-phonefilm.ca/ (accessed June 2016).

Tufte, T. (2005) *Media and Glocal Change: Rethinking Communication for Development*. Göteborg and Buenos Aires: Nordicom.

Turkle, S. (2011) *Alone Together: Why we Expect More From Technology and Less From Each Other*. New York: Basic Books.

Tuters, M. and Varnelis, K. (2011). Beyond locative media. *Networked Publics*. Available at: http://userwww.sfsu.edu/telarts/art511-locative/READINGS/beyondLocative.pdf. (accessed January 2016).

The Universal Declaration of Human Rights. Available at: http://www.un.org/en/documents/udhr/index.shtml (accessed October 2015).

*The Week* (2016) 'Spirit of the Age', July 16, p. 6.

Todorov, T. (1997) *The Poetics of Prose*. Ithaca, NY: New York University Press.

Tolle, E. (2001) *The Power of Now*. London: Holder and Stoughton.

Vaughan, D. (1999) *For Documentary: Twelve Essays*. Berkeley, CA: University of California Press.

Vertov, D. (1927) *In Kino-eye: The Writings of Dziga Vertov*. Berkeley: University of California Press.

Veruggio, G. (2008) Talking Robots: The podcast on Robotics and AI. Available at: http://lis2.epfl.ch/resources/podcast/2008/01/gianmarco-veruggio-roboethics.html (accessed October 2015).

VSCO-Cam. Available at: http://www.vsco-cam.com/ (accessed June 2016).

Wallace, S. (2009) Watchdog or Witness? The Emerging Forms and Practices of Videojournalism in *Journalism*, 10(5), 684–701.

Ward, M. (2013) Web porn: Just how much is there? Available at: http://www.bbc.co.uk/news/technology-23030090 (accessed January 2016).

Wardle, C. (2010) User generated content and public service broadcasting. *Networked Knowledge blog*, May 19. Available at: http://clairewardle.com/2010/05/19/user-generated-content-and-public-service-broadcasting/ (accessed January 2016).

Warner, M. (1994) *From the Beast to the Blond: On Fairy Tales and Their Tellers*. London: Chatto & Windus.

Watney, S. (1986) On the Institutions of Photography (pp. 141–161) in Evans, J. and Hall, S. (eds.) *Visual Culture: The Reader*. London: Sage Publications Ltd.

Wells, L. (2015) *Photography: A Critical Introduction London* (5th ed.). New York & London: Routledge.

Wells, N. (1910) A Description of the Affective Character of the Colors of the Spectrum in *The Psychological Bulletin*, V11(6), June 1910, 181–195. Available at: http://psycnet.apa.org/journals/bul/7/6/181 (accessed April 2016).

Whitehead, S. M. (2007, ed.) *Men and Masculinities*. Cambridge: Polity Press.

Williams, L. (1993) Mirrors without Memories: Truth, History and the New Documentary in *Film Quarterly*, 46 (3), 9–21.

Williams, V. (1991) T*he Other Observers: Women Photographers in Britain 1900 to the Present*. London: Virago Press.

Winter, D. (2011) Through my eye, not Hipstamatic's. Available at: http://lens.blogs.nytimes.com/2011/02/11/through-my-eye-not-hipstamatics/?_r=0 (accessed April 2016).

Wolff, J. (1981) *The Social Production of Art*. Basingstoke: Macmillan.

World Wide Web iRights, UK. Available at: http://irights.uk/ (accessed August 2015).

Wright, W. (1975) *Six Guns and Society: A Structural Study of the Western*. Berkley: University of California Press.

Wyatt, D. (2015) Art in Island: The art gallery designed for taking as many selfies as you want. Available at: http://www.independent.co.uk/arts-entertainment/art/features/art-in-island-the-art-gallery-designed-for-taking-as-many-selfies-as-you-want-10116247.html# (accessed July 2015).

YouGov (2016), YouGov on the day poll: Remain 52%, Leave 48%. Available at: https://yougov.co.uk/news/2016/06/23/yougov-day-poll (accessed July 17 2016).

Yuhas, A. (2015) Man who filmed Walter Scott shooting: I worry what might happen to me. Available at: http://www.theguardian.com/us-news/2015/apr/09/walter-scott-shooting-south-carolina-feidin-santana (accessed October 2015).

Zamora, L. and Faris, W. B. (1995) *Magical Realism: Theory, History, Community*. Durham, NC and London: Duke University Press.

Zipes, J. (1993) *The Trials and Tribulations of Little Red Riding Hood*. New York & London: Routledge.

Zylinska, J. (2014) *Minimal Ethics for the Anthropocene*. Michigan: Open Humanities Press.

# Index

www.ingramcontent.com/pod-product-compliance
Lightning Source LLC
Chambersburg PA
CBHW060237220326
41598CB00027B/3960